PROBLEM-SOLVING CASES IN MICROSOFT® ACCESS AND EXCEL, FIFTH ANNUAL EDITION

Joseph A. Brady

Ellen F. Monk

THOMSON

COURSE TECHNOLOGY

Australia • Canada • Mexico • Singapore • Spain • United Kingdom • United States

D1402531

THOMSON
COURSE TECHNOLOGY

Problem-Solving Cases in Microsoft® Access and Excel,
Fifth Annual Edition

Acquisitions Editor:
Maureen Martin

Product Managers:
Kate Hennessy
Tricia Coia

Development Editors:
Saher Alam
DeVona Dors

Marketing Specialist:
Vicki Ortiz

Editorial Assistant:
Erin Kennedy

Content Product Manager:
Catherine G. DiMassa
GEX Publishing Services

Manufacturing Coordinator:
Justin Palmeiro

Cover Designer:
Beth Paquin

Compositor:
GEX Publishing Services

Dedication

To my wife, Karen—a great problem-solver
JAB
To my sisters, Lauren and Joan
EFM

For nearly two decades, we have taught MIS courses at the University of Delaware. From the start, we wanted to use good computer-based case studies for the database and decision-support portions of our courses.

We could not find a casebook that met our needs! This surprised us because our requirements, we thought, were not unreasonable. First, we wanted cases that asked students to think about real-world business situations. Second, we wanted cases that provided students with hands-on experience, using the kind of software that they had learned to use in their computer literacy courses—and that they would later use in business. Third, we wanted cases that would strengthen students' ability to analyze a problem, examine alternative solutions, and implement a solution using software. Undeterred by the lack of casebooks, we wrote our own cases, and Thomson Course Technology published them.

This book is the sixth casebook we have written for Thomson Course Technology. The cases are all new and the tutorials are updated. New features of this textbook reflect comments and suggestions that we have received from instructors who have used our casebook in their classrooms. In addition, the current cases have been written for the new, Office 2007 versions of Microsoft Access and Excel.

As with our prior casebooks, we include tutorials that prepare students for the cases, which are challenging but doable. Most of the cases are organized in a way that helps students think about the logic of each case's business problem and then about how to use the software to solve the business problem. The cases fit well in an undergraduate MIS course, an MBA Information Systems course, or a Computer Science course devoted to business-oriented programming.

BOOK ORGANIZATION

The book is organized into seven parts:

1. Database Cases Using Access

2. Decision Support Cases Using Excel Scenario Manager

3. Decision Support Cases Using the Excel Solver

4. Decision Support Case Using Basic Excel Functionality

5. Integration Case: Using Access and Excel

6. Advanced Excel Skills Using Excel

7. Presentation Skills

Part 1 begins with two tutorials that prepare students for the Access case studies. Parts 2 and 3 each begin with a tutorial that prepares students for the Excel case studies. All four tutorials provide students with hands-on practice in using the software's more advanced features—the kind of support that other books about Access and Excel do not give to students. Part 4 asks students to use Excel's basic functionality for decision support. Part 5 challenges students to use both Access and Excel to find a solution to solve a business problem. Part 6 is a set of short tutorials on the advanced skills students need to complete some of the Excel cases. Part 7 is a tutorial that hones students' skills in creating and delivering an oral presentation to business managers. The next section explores each of these parts in more depth.

Part 1: Database Cases
Using Access

This section begins with two tutorials and then presents five case studies.

Tutorial A: Database Design

This tutorial helps students to understand how to set up tables to create a database, without requiring students to learn formal analysis and design methods, such as data normalization.

Tutorial B: Microsoft Access

The second tutorial teaches students the more advanced features of Access queries and reports—features that students will need to know to complete the cases.

Cases 1–5

Five database cases follow Tutorials A and B. The students' job is to implement each case's database in Access so form, query, navigation pane, and report outputs can help management. The first case is an easier "warm-up" case. The next four cases require a more demanding database design and implementation effort.

Part 2: Decision Support Cases
Using the Excel Scenario Manager

This section has one tutorial and two decision support cases requiring the use of Excel Scenario Manager.

Tutorial C: Building a Decision Support System in Excel

This section begins with a tutorial using Excel for decision support and spreadsheet design. Fundamental spreadsheet design concepts are taught. Instruction on the Scenario Manager, which can be used to organize the output of many "what-if" scenarios, is emphasized.

Cases 6–7

These two cases can be done with or without the Scenario Manager (although the Scenario Manager is nicely suited to them). In each case, students must use Excel to model two or more solutions to a problem. Students then use the outputs of the model to identify and document the preferred solution via a memorandum and, if assigned to do so, an oral presentation.

Part 3: Decision Support Cases
Using the Excel Solver

This section has one tutorial and two decision support cases requiring the use of Excel Solver.

Tutorial D: Building a Decision Support System Using Excel Solver

This section begins with a tutorial about using the Solver, which is a decision support tool for solving optimization problems.

Cases 8–9

Once again, in each case, students use Excel to analyze alternatives and identify and document the preferred solution.

Part 4: Decision Support Cases
Using Basic Excel Functionality

Case 10

The cases continue with one case that uses basic Excel functionality, i.e., the case does not require the Scenario Manager or the Solver. Excel is used to test students' analytical skills in "what if" analyses.

Part 5: Integration Case
Using Excel and Access

Case 11

This case integrates Access and Excel. This case is included because of a trend toward sharing data among multiple software packages to solve problems.

Part 6: Advanced Excel Skills Using Excel

Using Excel

This part has one tutorial forcused on advanced techniques in Excel.

Tutorial E: Guidance for Excel Cases

A number of the cases in this book require the use of some advanced techniques in Excel. The following techniques are explained in this tutorial (rather than in the cases themselves): VLOOKUP function, PMT function, CUMPRINC function, and Data tables.

Part 7: Presentation Skills

Tutorial F: Giving an Oral Presentation

Each case includes an optional presentation assignment that gives students practice in making a presentation to management on the results of their analysis of the case. This section gives advice on how to create oral presentations. It also has technical information on charting and pivot tables, techniques that might be useful in case analyses or as support for presentations. This tutorial will help students to organize their recommendations, to present their solutions in both words and graphics, and to answer questions from the audience. For larger classes, instructors may wish to have students work in teams to create and deliver their presentations—which would model the "team" approach used by many corporations.

To view and access additional cases, instructors should see the "Hall of Fame" note in the *Using the Cases* section shown below.

INDIVIDUAL CASE DESIGN

The format of the eleven cases follows this template.

- Each case begins with a *Preview* of what the case is about and an overview of the tasks.

- The next section, *Preparation*, tells students what they need to do or know to complete the case successfully. (Of course, our tutorials prepare students for the cases!)

- The third section, *Background*, provides the business context that frames the case. The background of each case models situations that require the kinds of thinking and analysis that students will need in the business world.

- The Background sections are followed by the *Assignment* sections, which are generally organized in a way that helps students to develop their analyses.

- The last section, *Deliverables*, lists what students must hand in: printouts, a memorandum, a presentation, and files on disk. The list is similar to the kind of deliverables that a business manager might demand.

USING THE CASES

We have successfully used cases like these in our undergraduate MIS courses. We usually begin the semester with Access database instruction. We assign the Access database tutorials and then a case to each student. Then, for Excel DSS instruction, we do the same thing—assign a tutorial and then a case.

Some instructors have expressed an interest in having access to extra cases, especially in the second semester of a school year.For example: "I assigned the integration case in the fall, and need another one for the spring." To meet this need, we have set up an online "Hall of Fame" that features some of our favorite cases from prior editions. This password-protected Hall of Fame is available to instructors and can be found on the Thomson Course Technology Web site. Go to *www.course.com* and search for this textbook by title, author, or ISBN. Note that the cases are in MS Office 2003 format, but MS Office 2007 will read and tranlate them easily.

TECHNICAL INFORMATION

This textbook was quality assurance tested using the Windows XP Professional operating system, Microsoft Access 2007, and Microsoft Excel 2007.

Data Files and Solution Files

We have created "starter" data files for the Excel cases, so students need not spend time typing in the spreadsheet skeleton. Case 11 also requires students to load an Access database file. All these files can be found on the Thomson Course Technology Web site, which is available to both students and instructors. Go to *www.course.com* and search for this textbook by title, author, or ISBN. You are granted a license to copy the data files to any computer or computer network used by individuals who have purchased this textbook.

Solutions to the material in the text are available to instructors. These can also be found at *www.course.com*. Search for this textbook by title, author, or ISBN. The solutions are password protected.

Instructor's Manual

An Instructor's Manual is available to accompany this text. The Instructor's Manual contains additional tools and information to help instructors successfully use this textbook. Items such as a Sample Syllabus, Teaching Tips, and Grading Guidelines can be found in the Instructor's Manual. Instructors should go to *www.course.com* and search for this textbook by title, author, or ISBN. The Instructor's Manual is password protected.

ACKNOWLEDGEMENTS

We would like to give many thanks to the team at Thomson Course Technology, including our Developmental Editors, Saher Alam and DeVona Dors; Product Managers, Kate Hennessy and Tricia Coia; and to our Content Project Managers, Sandra Mitchell and Cathie DiMassa. As always, we acknowledge our students' diligent work.

TABLE OF CONTENTS

Part 5: Integration Case: Using Access and Excel

Part 6:   Advanced Excel Skills Using Excel

Part 7:   Presentation Skills

PART 1

DATABASE CASES
USING ACCESS

TUTORIAL **A**

DATABASE DESIGN

This tutorial has three sections. The first section briefly reviews basic database terminology. The second section teaches database design. The third section features a practice database design problem.

REVIEW OF TERMINOLOGY

You will begin by reviewing some basic terms that will be used throughout this textbook. In Access, a **database** is a group of related objects that are saved in one file. An Access **object** can be a table, a form, a query, or a report. You can identify an Access database file by its suffix .accdb.

A **table** consists of data that is arrayed in rows and columns. A **row** of data is called a **record**. A **column** of data is called a **field**. Thus, a record is a set of related fields. The fields in a table should be related to one another in some way. For example, a company might want to keep its employee data together by creating a database table called EMPLOYEE. That table would contain data fields about employees, such as their names and addresses. It would not have data fields about the company's customers; that data would go in a CUSTOMER table.

A field's values have a **data type** that is declared when a table is defined. That way, when data is entered into the database, the software knows how to interpret each entry. The data types found in Access include the following:

- "Text" for words
- "Integer" for whole numbers
- "Double" for numbers that have a decimal value
- "Currency" for numbers that should be treated as dollars and cents
- "Yes/No" for variables that have only two values (1-0, on/off, yes/no, true/false)
- "Date/Time" for variables that are dates or times

Each database table should have a **primary key** field, a field in which each record has a *unique* value. For example, in an EMPLOYEE table, a field called SSN (for Social Security number) could serve as a primary key, because each record's SSN value would be different from every other record's SSN value. Sometimes a table does not have a single field whose values are all different. In that case, two or more fields are combined into a **compound primary key**. The combination of the fields' values is unique.

Database tables should be logically related to one another. For example, suppose a company has an EMPLOYEE table with fields for SSN, Name, Address, and Telephone Number. For payroll purposes, the company has an HOURS WORKED table with a field that summarizes Labor Hours for individual employees. The relationship between the EMPLOYEE table and the HOURS WORKED table needs to be established in the database so you can tell which employees worked which hours. That is done by including the primary key field from the EMPLOYEE table (SSN) as a field in the HOURS WORKED table. In the HOURS WORKED table, the SSN field is then called a **foreign key**.

In Access, data can be entered directly into a table or it can be entered into a form, which then inserts the data into a table. A **form** is a database object that is created from an existing table to make the process of entering data more user-friendly.

A **query** is the database equivalent of a question that is posed about data in a table (or tables). For example, suppose a manager wants to know the names of employees who have worked for the company for more than five years. A query could be designed in such a way that it interrogates the EMPLOYEE table to search for the information. The query would be run, and its output would answer the question.

Because a query may need to pull data from more than one table, queries can be designed to interrogate more than one table at a time. In that case, the tables must be connected by a **join** operation, which links tables on the values in a field that they have in common. The common field acts as a "hinge" for the joined tables; and when the query is run, the query generator treats the joined tables as one large table.

In Access, queries that answer a question are called **select** queries, because they select relevant data from the database records. Queries also can be designed to change data in records, add a record to the end of a table, or delete entire records from a table. Those are called **update**, **append**, and **delete** queries, respectively.

Access has a **report** generator that can be used to format a table's data or a query's output.

DATABASE DESIGN

Designing a database involves determining which tables need to be in the database and creating the fields that need to be in each table. This section begins with an introduction to key database design concepts, then discusses database design rules. Those rules are a series of steps you should use when building a database. First, the following key concepts are defined:

- Entities
- Relationships
- Attributes

Database Design Concepts

Computer scientists have highly formalized ways of documenting a database's logic, but learning their notations and mechanics can be time-consuming and difficult. In fact, doing so usually takes a good portion of a systems analysis and design course. This tutorial will teach you database design by emphasizing practical business knowledge, and the approach should enable you to design serviceable databases. Your instructor may add more formal techniques.

A database models the logic of an organization's operation, so your first task is to understand that operation. You do that by talking to managers and workers, by making observations, and/or by looking at business documents such as sales records. Your goal is to identify the business's "entities" (sometimes called *objects*). An **entity** is some thing or some event that the database will contain. Every entity has characteristics, called **attributes**, and a **relationship** (or relationships) to other entities. Take a closer look.

Entities

As previously mentioned, an entity is a tangible thing or an event. The reason for identifying entities is that *an entity eventually becomes a table in the database*. Entities that are things are easy to identify. For example, consider a video store. The database for the video store would probably need to contain the names of DVDs and the names of customers who rent them, so you would have one entity named VIDEO and another named CUSTOMER.

In contrast, entities that are events can be more difficult to identify, probably because although events cannot be seen, they are no less real. In the video store example, one event would be VIDEO RENTAL and another event would be HOURS WORKED by employees.

In general, your analysis of an organization's operations can be made easier by knowing that organizations usually have certain physical entities such as these:

- Employees
- Customers
- Inventory (products or services)
- Suppliers

Thus, the database for most organizations would have a table for each of those entities. Your analysis also can be made easier by knowing that organizations engage in transactions internally (within the company) and externally (with the outside world). Those transactions are the subject of any accounting course, but most people understand them from events that occur in daily life. Consider the following examples:

- Organizations generate revenue from sales or interest earned. Revenue-generating transactions include event entities called SALES and INTEREST.
- Organizations incur expenses from paying hourly employees and purchasing materials from suppliers. HOURS WORKED and PURCHASES are event entities in the databases of most organizations.

Thus, identifying entities is a matter of observing what happens in an organization. Your powers of observation are aided by knowing what entities exist in the databases of most organizations.

Relationships

As a database analyst building a database, you should consider the relationship of each entity to other entities. For each entity, you should ask, "What is the relationship, if any, of this entity to every other entity identified?" Relationships can be expressed in English. For example, suppose a college's database has entities for STUDENT (containing data about each student), COURSE (containing data about each course), and SECTION (containing data about each section). A relationship between STUDENT and SECTION would be expressed as "Students enroll in sections."

An analyst also must consider the **cardinality** of any relationship. Cardinality can be one-to-one, one-to-many, or many-to-many. Those relationships are summarized as follows:

- In a one-to-one relationship, one instance of the first entity is related to just one instance of the second entity.
- In a one-to-many relationship, one instance of the first entity is related to many instances of the second entity; but each instance of the second entity is related to only one instance of the first entity.
- In a many-to-many relationship, one instance of the first entity is related to many instances of the second entity and one instance of the second entity is related to many instances of the first entity.

For a more concrete understanding of cardinality, consider again the college database with the STUDENT, COURSE, and SECTION entities. The university catalog shows that a course such as Accounting 101 can have more than one section: 01, 02, 03, 04, etc. Thus, the following relationships can be observed:

- The relationship between the entities COURSE and SECTION is one-to-many. Each course has many sections, but each section is associated with just one course.
- The relationship between STUDENT and SECTION is many-to-many. Each student can be in more than one section, because each student can take more than one course. Also, each section has more than one student.

Thinking about relationships and their cardinalities may seem tedious to you. But as you work through the cases in this text, you will see that this type of analysis and the knowledge it yields can be very valuable in designing databases. In the case of many-to-many relationships, you should determine the database tables a given database needs; and in the case of one-to-many relationships, you should decide which fields the database's tables need to share.

Attributes

An attribute is a characteristic of an entity. The reason you identify attributes of an entity is because *attributes become a table's fields*. If an entity can be thought of as a noun, an attribute can be thought of as an adjective describing the noun. Continuing with the college database example, consider the STUDENT entity. Students have names. Thus, Last Name would be an attribute of the entity called STUDENT and, therefore, a field in the STUDENT table. First Name would be an attribute as well. The STUDENT entity also would have an Address attribute as another field, along with Phone Number and whatever other fields were used.

Sometimes it can be difficult to tell the difference between an attribute and an entity. One good way to differentiate them is to ask whether more than one attribute is possible for each entity. If more than one instance is possible and you do not know in advance how many there will be, then it's an entity. For example, assume a student could have two (but no more than two) Addresses—one for home and one for college. You could specify attributes Address 1 and Address 2. Now consider what would happen if the number of student addresses could not be stipulated in advance, meaning all addresses had to be recorded. In that case, you would not know how many fields to set aside in the STUDENT table for addresses. Therefore, you would need a separate STUDENT ADDRESSES table (entity) that would show any number of addresses for a given student.

DATABASE DESIGN RULES

As described previously, your first task in database design is to understand the logic of the business situation. Once you understand that, you are ready to build a database for the requirements of the situation. To create a context for learning about database design, look at a hypothetical business operation and its database needs.

Example: The Talent Agency

Suppose you have been asked to build a database for a talent agency. The agency books bands into nightclubs. The agent needs a database to keep track of the agency's transactions and to answer day-to-day questions.

Many questions arise in the running of this business. For example, a club manager often wants to know which bands are available on a certain date at a certain time or wants to know the agent's fee for a certain band. Similarly, the agent may want to see a list of all band members and the instrument each person plays or a list of all bands having three members.

Suppose you have talked to the agent and have observed the agency's business operation. You conclude that your database needs to reflect the following facts:

1. A booking is an event in which a certain band plays in a particular club on a particular date, starting at a certain time, ending at a certain time, and performing for a specific fee. A band can play more than once a day. The Heartbreakers, for example, could play at the East End Cafe in the afternoon and then at the West End Cafe that same night. For each booking, the club pays the talent agent. The agent keeps a 5 percent fee and then gives the remainder of the payment to the band.

2. Each band has at least two members and an unlimited maximum number of members. The agent notes a telephone number of just one band member, which is used as the band's contact number. No two bands have the same name or telephone number.

3. No members of any of the bands have the same name. For example, if there is a Sally Smith in one band, there is no Sally Smith in another band.

4. The agent keeps track of just one instrument that each band member plays. For this record keeping purpose, "vocals" are considered an instrument.

5. Each band has a desired fee. For example, the Lightmetal band might want $700 per booking and would expect the agent to try to get at least that amount.

6. Each nightclub has a name, an address, and a contact person. That person has a telephone number that the agent uses to contact the club. No two clubs have the same name, contact person, or telephone number. Each club has a target fee. The contact person will try to get the agent to accept that amount for a band's appearance.

7. Some clubs feed the band members for free; others do not.

Before continuing with this tutorial, you might try to design the agency's database on your own. Ask yourself, What are the entities? Recall that databases usually have CUSTOMER, EMPLOYEE, and INVENTORY entities as well as an entity for the revenue-generating transaction event. Each entity becomes a table in the database. What are the relationships between entities? For each entity, what are its attributes? The answers to those questions become the fields in each table. For each table, what is the primary key?

Six Database Design Rules

Assume you have gathered information about the business situation in the talent agency example. Now you want to identify the tables required for the database and the fields needed in each table. To do that, observe the following six rules:

Rule 1: You do not need a table for the business. The database represents the entire business. Thus, in the example, Agent and Agency are not entities.

Rule 2: Identify the entities in the business description. Look for the things and events that the database must contain. Those become tables in the database. Typically, certain entities are represented. In the talent agency example, you should be able to observe these entities:

- *Things*: The product (inventory for sale) is Band. The customer is Club.
- *Events*: The revenue-generating transaction is Bookings.

You might ask yourself, Is there an EMPLOYEE entity? Isn't INSTRUMENT an entity? Those issues will be discussed as the rules are explained.

Rule 3: Look for relationships between the entities. Look for one-to-many relationships between entities. The relationship between those entities must be established in the tables, and that is done by using a foreign key. Those mechanics are explained in the next rule's discussion of the relationship between Band and Band Member.

Look for many-to-many relationships between entities. In each of those relationships, there is the need for a third entity that associates the two entities in the relationship. Recall from the college database scenario the many-to-many relationship example that involved STUDENT and SECTION entities. To show the ENROLLMENT of specific students in specific sections, a third table needs to be created. The mechanics of doing that are described in the next rule in the discussion of the relationship between BAND and CLUB.

Rule 4: Look for attributes of each entity and designate a primary key. As previously mentioned, you should think of the entities in your database as nouns. You should then create a list of adjectives that describe those "nouns." Those adjectives are the attributes that will become the table's fields. After you have identified fields for each table, you should check to see if a field has unique values. If one exists, designate it as the primary key field; otherwise, designate a compound primary key.

Returning to the talent agency example, the attributes, or fields, of the BAND entity are Band Name, Band Phone Number, and Desired Fee. No two bands have the same names, so the primary key field can be Band Name. Figure A-1 shows the BAND table and its fields: Band Name, Band Phone Number, and Desired Fee; the data type of each field also is shown.

BAND	
Field	*Data Type*
Band Name (primary key)	Text
Band Phone Number	Text
Desired Fee	Currency

FIGURE A-1   The BAND table and its fields

Two BAND records are shown in Figure A-2.

Band Name (primary key)	Band Phone Number	Desired Fee
Heartbreakers	981 831 1765	$800
Lightmetal	981 831 2000	$700

FIGURE A-2   Records in the BAND table

If two bands could have the same name (that is, if uniqueness in band names wasn't assumed), Band Name would not be a good primary key and some other unique identifier would be needed. Those situations are common. Most businesses have many types of inventory, and duplicate names are possible. The typical solution is to assign a number to each product to be used as the primary key field. For example, a college could have more than one faculty member with the same name, so each faculty member would be assigned an employee identification number (EIN). Similarly, banks assign a personal identification number (PIN) for each depositor. Each automobile that a car manufacturer makes gets a unique Vehicle Identification Number (VIN). Most businesses assign a number to each sale, called an invoice number. (The next time you buy something at a grocery store, note the number on your receipt. It will be different from the number the next person in line sees on his or her receipt.)

At this point, you might be wondering why Band Member would not be an attribute of BAND. The answer is that although you must record each band member, you do not know in advance how many members will be in each band. Therefore, you do not know how many fields to allocate to the BAND table for members. Another way to think about Band Member(s) is that they are, in effect, the agency's employees. Databases for organizations usually have an EMPLOYEE entity. Therefore, you should create a BAND MEMBER table with the attributes Member Name, Band Name, Instrument, and Phone. The BAND MEMBER table and its fields are shown in Figure A-3.

BAND MEMBER	
Field Name	*Data Type*
Member Name (primary key)	Text
Band Name (foreign key)	Text
Instrument	Text
Phone	Text

FIGURE A-3   The BAND MEMBER table and its fields

Note in Figure A-3 that the phone number is classified as a text data type. The data type for such "numbers" is text—and not number—because the values will not be used in an arithmetic computation. The benefit is that text data type values take up fewer bytes than numerical or currency data type values; therefore, the file uses less storage space. That rule of using text data types would also hold for values such as ZIP Codes, Social Security numbers, etc.

Five records in the BAND MEMBER table are shown in Figure A-4 section.

Member Name (primary key)	Band Name	Instrument	Phone
Pete Goff	Heartbreakers	Guitar	981 444 1111
Joe Goff	Heartbreakers	Vocals	981 444 1234
Sue Smith	Heartbreakers	Keyboard	981 555 1199
Joe Jackson	Lightmetal	Sax	981 888 1654
Sue Hoopes	Lightmetal	Piano	981 888 1765

FIGURE A-4   Records in the BAND MEMBER table

Instrument can be included as a field in the BAND MEMBER table, because the agent records only one instrument for each band member. Thus, instrument can be thought of as a way to describe a band member, much like the phone number is part of the description. Member Name can be the primary key because of the assumption (albeit arbitrary) that no two members in any band have the same name. Alternatively, Phone could be the primary key, assuming no two members share a telephone. Or a band member ID number could be assigned to each person in each band, which would create a unique identifier for each band member that the agency handled.

You might ask why Band Name is included in the BAND MEMBER table. The commonsense reason is that you did not include the Member Name in the BAND table. You must relate bands and members somewhere, and BAND table is the place to do it.

Another way to think about this involves the cardinality of the relationship between BAND and BAND MEMBER. It is a one-to-many relationship: one band has many members, but each member is in just one band. You establish that kind of relationship in the database by using the primary key field of one table as a foreign key in the other table. In BAND MEMBER, the foreign key Band Name is used to establish the relationship between the member and his or her band.

The attributes of the entity CLUB are Club Name, Address, Contact Name, Club Phone Number, Preferred Fee, and Feed Band? The table called CLUB can define the CLUB entity, as shown in Figure A-5.

CLUB	
Field Name	*Data Type*
Club Name (primary key)	Text
Address	Text
Contact Name	Text
Club Phone Number	Text
Preferred Fee	Currency
Feed Band?	Yes/No

FIGURE A-5   The CLUB table and its fields

Two records in the CLUB table are shown in Figure A-6.

Club Name (primary key)	Address Name	Contact Number	Club Phone	Preferred Fee	Feed Band?
East End	1 Duce St.	Al Pots	981 444 8877	$600	Yes
West End	99 Duce St.	Val Dots	981 555 0011	$650	No

FIGURE A-6   Records in the CLUB table

You might wonder why Bands Booked into Club (or some such field name) is not an attribute of the CLUB table. There are two reasons. First, because you do not know in advance how many bookings a club will have, the value cannot be an attribute. Second, BOOKINGS is the agency's revenue-generating transaction, an event entity, and you need a table for that business transaction. Consider the booking transaction next.

You know that the talent agent books a certain band into a certain club on a certain date for a certain fee, starting and ending at a certain time. From that information, you can see that the attributes of the BOOKINGS entity are Band Name, Club Name, Date, Start Time, End Time, and Fee. The BOOKINGS table and its fields are shown in Figure A-7.

BOOKINGS	
Field Name	*Data Type*
Band Name	Text
Club Name	Text
Date	Date/Time
Start Time	Date/Time
End Time	Date/Time
Fee	Currency

FIGURE A-7   The BOOKINGS table and its fields—and no designation of a primary key

Some records in the BOOKINGS table are shown in Figure A-8.

Band Name	Club Name	Date	Start Time	End Time	Fee
Heartbreakers	East End	11/21/08	19:00	23:30	$800
Heartbreakers	East End	11/22/08	19:00	23:30	$750
Heartbreakers	West End	11/28/08	13:00	18:00	$500
Lightmetal	East End	11/21/08	13:00	18:00	$700
Lightmetal	West End	11/22/08	13:00	18:00	$750

FIGURE A-8   Records in the BOOKINGS table

Note that no single field is guaranteed to have unique values, because each band is likely to be booked many times and each club used many times. Further, each date and time also can appear more than once. Thus, no one field can be the primary key.

If a table does not have a single primary key field, you can make a compound primary key whose field values, when taken together, will be unique. Because a band can be in only one place at a time, one possible solution is to create a compound key consisting of the fields Band Name, Date, and Start Time. An alternative solution is to create a compound primary key consisting of the fields Club Name, Date, and Start Time.

A way to avoid having a compound key is to create a field called Booking Number. Each booking would have its own unique number, similar to an invoice number.

Here is another way to think about this event entity: Over time, a band plays in many clubs and each club hires many bands. Thus, the BAND-to-CLUB relationship is a many-to-many relationship. Such relationships signal the need for a table between the two entities in the relationship. Here you would need the BOOKINGS table, which associates the BAND and CLUB tables. An associative table is implemented by including the primary keys from the two tables that are associated. In this case, the primary keys from the BAND and CLUB tables are included as foreign keys in the BOOKINGS table.

Rule 5: Avoid data redundancy. You should not include extra (redundant) fields in a table. Doing so takes up extra disk space and leads to data entry errors because the same value must be entered in multiple tables, increasing the chance of a keystroke error. In large databases, keeping track of multiple instances of the same data is nearly impossible and contradictory data entries become a problem.

Consider this example: Why wouldn't Club Phone Number be in the BOOKINGS table as a field? After all, the agent might have to call about some last-minute change for a booking and could quickly look up the number in the BOOKINGS table. Assume the BOOKINGS table had Booking Number as the primary key and Club Phone Number as a field. Figure A-9 shows the BOOKINGS table with the additional field.

BOOKINGS	
Field Name	*Data Type*
Booking Number (primary key)	Text
Band Name	Text
Club Name	Text
Club Phone Number	Text
Date	Date/Time
Start Time	Date/Time
End Time	Date/Time
Fee	Currency

FIGURE A-9   The BOOKINGS table with an unnecessary field—Club Phone Number

The fields Date, Start Time, End Time, and Fee logically depend on the Booking Number primary key—they help define the booking. Band Name and Club Name are foreign keys and are needed to establish the relationship between the tables BAND, CLUB, and BOOKINGS. But what about Club Phone Number? It is not defined by the Booking Number. It is defined by Club Name—*that is, it's a function of the club, not of the booking.* Thus, the Club Phone Number field does not belong in the BOOKINGS table. It's already in the CLUB table; and if the agent needs the Club Phone Number field, he or she can look it up there.

Perhaps you can see the practical data entry problem of including Club Phone Number in BOOKINGS. Suppose a club changed its contact phone number. The agent can easily change the number one time, in CLUB. But now the agent would need to remember the names of all of the other tables that had that field and change the values there too. Of course, with a small database, that might not be difficult. But in large databases, having many redundant fields in many tables makes this sort of maintenance very difficult, which means that redundant data is often incorrect.

You might object, saying, "What about all of those foreign keys? Aren't they redundant?" In a sense, they are. But they are needed to establish the relationship between one entity and another, as discussed previously.

Rule 6: Do not include a field if it can be calculated from other fields. A calculated field is made using the query generator. Thus, the agent's fee is not included in the BOOKINGS table, because it can be calculated by query (here, 5 percent times the booking fee).

PRACTICE DATABASE DESIGN PROBLEM

Imagine this scenario: Your town has a library that wants to keep track of its business in a database, and you have been called in to build the database. You talk to the town librarian, review the old paper-based records, and watch people use the library for a few days. You learn these things about the library:

1. Anyone who lives in the town can get a library card if he or she asks for one. The library considers each person who gets a card a "member" of the library.

2. The librarian wants to be able to contact members by telephone and by mail. She calls members when their books are overdue or when requested materials become available. She likes to mail a thank-you note to each patron on his or her yearly anniversary of becoming a member of the library. Without a database, contacting members can be difficult to do efficiently; for example, there could be more than one member by the name of Martha Jones. Also, a parent and a child often have the same first and last name, live at the same address, and share a phone.

3. The librarian tries to keep track of each member's reading interests. When new books come in, the librarian alerts members whose interests match those books. For example, long-time member Sue Doaks is interested in reading Western novels, growing orchids, and baking bread. There must be some way to match her interests with available books. One complication is that although the librarian wants to track all of a member's reading interests, she wants to classify each book as being in just one category of interest. For example, the classic gardening book *Orchids of France* would be classified as a book about orchids or a book about France, but not both.

4. The library stocks many books. Each book has a title and any number of authors. Also, there could, conceivably, be more than one book in the library titled *History of the United States*. Similarly, there could be more than one author with the same name.

5. A writer could be the author of more than one book.

6. A book could be checked out repeatedly as time goes on. For example, *Orchids of France* could be checked out by one member in March, another member in July, and yet another member in September.

7. The library must be able to identify whether a book is checked out.

8. A member can check out any number of books in a visit. It's also conceivable that a member could visit the library more than once a day to check out books—some members do.

9. All books that are checked out are due back in two weeks, no exceptions. The late fee is 50 cents per day late. The librarian would like to have an automated way of generating an overdue book list each day so she can telephone the miscreants.

10. The library has a number of employees. Each employee has a job title. The librarian is paid a salary, but other employees are paid by the hour. Employees clock in and out each day. Assume all employees work only one shift per day and all are paid weekly. Pay is deposited directly into an employee's checking account—no checks are hand-delivered. The database needs to include the librarian and all other employees.

Design the library's database, following the rules set forth in this tutorial. Your instructor will specify the format of your work. Here are a few hints in the form of questions:

- A book can have more than one author. An author can write more than one book. How would you describe the relationship between books and authors?

- The library lends books for free, of course. If you were to think of checking out a book as a sale transaction for zero revenue, how would you handle the library's revenue-generating event?

- A member can borrow any number of books at a checkout. A book can be checked out more than once. How would you describe the relationship between checkouts and books?

MICROSOFT ACCESS TUTORIAL

Microsoft Access is a relational database package that runs on the Microsoft Windows operating system. This tutorial was prepared using Access 2007.

Before using this tutorial, you should be familiar with the fundamentals of Microsoft Access and know how to use Windows. This tutorial teaches you some of the advanced Access skills you'll need to do database case studies. The tutorial concludes with a discussion of common Access problems and explains how to solve them.

A preliminary word of caution: always observe proper file-saving and closing procedures. Use these steps to exit Access: (1) Office Button—Close Database, then (2) X-Exit Access Button. Or you may simply click the **X-Exit Access** Button, which gets you back to Windows. Always end your work with those file-closing steps. Do not remove your disk, CD, or other portable storage device such as a USB flash drive when database forms, tables, etc., appear on the screen; you will lose your work.

To begin this tutorial, you will create a new database called **Employee**.

AT THE KEYBOARD

Open a new database. On the Getting Started with Microsoft Office Access page, under the heading New Blank Database, click Blank Database. Call the database **Employee**. Click the file folder next to the filename to browse for the desired folder. Otherwise, your file will be saved automatically in My Documents.

Your opening screen should resemble the screen shown in Figure B-1.

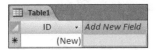

FIGURE B-1   Entering data in the Datasheet view

When you create a table, Access opens the table in the Datasheet view by default. Since you will be using the Design View to build your tables, to close the new table, click the *x* in the upper right-hand corner of the table window that corresponds to Close Table I. You are now on the Home tab in the Database window of Access, as shown in Figure B-2. From this screen, you can create or change objects.

FIGURE B-2   The Database window Home tab in Access

CREATING TABLES

Your database will contain data about employees, their wage rates, and the hours they worked.

Defining Tables

In the Database window, build three new tables using the instructions that follow.

(1) Define a table called EMPLOYEE.

This table contains permanent data about employees. To create it, choose the Create tab and in the Tables group, click **Table design**. The table's fields are Last Name, First Name, SSN (Social Security Number), Street Address, City, State, Zip, Date Hired, and US Citizen. The field SSN is the primary key field. Change the length of text fields from the default 255 spaces to more appropriate lengths; for example, the field Last Name might be 30 spaces and the Zip field might be 10 spaces. Your completed definition should resemble the one shown in Figure B-3.

Field Name	Data Type	Description
Last Name	Text	
First Name	Text	
SSN	Text	
Street Address	Text	
City	Text	
State	Text	
Zip	Text	
Date Hired	Date/Time	
US Citizen	Yes/No	

FIGURE B-3   Fields in the EMPLOYEE table

When you're finished, choose Office Button—Save; then enter the name desired for the table (here, EMPLOYEE). Make sure you specify the name of the *table*, not the database itself. (Here, it is a coincidence that the EMPLOYEE table has the same name as its database file.) Close the table by clicking the x-Close corresponding to the table EMPLOYEE.

(2) Define a table called WAGE DATA.

This table contains permanent data about employees and their wage rates. The table's fields are SSN, Wage Rate, and Salaried. The field SSN is the primary key field. Use the data types shown in Figure B-4. Your definition should resemble the one shown in Figure B-4.

Field Name	Data Type	Description
SSN	Text	
Wage Rate	Currency	
Salaried	Yes/No	

FIGURE B-4   Fields in the WAGE DATA table

Use Office Button—Save to save the table definition. Name the table Wage Data.

(3) Define a table called HOURS WORKED.

The purpose of this table is to record the number of hours employees work each week during the year. The table's fields are SSN (text), Week # (number—long integer), and Hours (number—double). The SSN and Week # are the compound keys.

In the following example, the employee with SSN 089-65-9000 worked 40 hours in Week 1 of the year and 52 hours in Week 2.

SSN	Week #	Hours
089-65-9000	1	40
089-65-9000	2	52

Note that no single field can be the primary key field. Why? The reason is that 089-65-9000 is an entry for each week. In other words, if this employee works each week of the year, at the end of the year, 52 records will have the SSN value. Thus, SSN values will not distinguish records. In addition, no other single field can distinguish these records either, because other employees will have worked during the same week number and some employees will have worked the same number of hours. For example, 40 hours—corresponding to a full-time workweek—would be a common entry for many weeks.

However, that presents a problem because in Access, a table must have a primary key field. What is the solution? Use a compound primary key; that is, use values from more than one field to create a combined field that will distinguish records. The best compound key to use for the current example consists of the field SSN and the Week # field. Why? Because as each person works each week, the week passes. That means, for example, that there is only *one* combination of SSN 089-65-9000 and Week # 1. Because those values *can occur in only one record*, the combination distinguishes that record from all others.

How do you set a compound key? The first step is to highlight the fields in the key. Those fields must appear one after the other in the table definition screen. (Plan ahead for that format.) As an alternative, you can highlight one field, hold down the Control key, and highlight the next field.

AT THE KEYBOARD

For the HOURS WORKED table, click the first field's left prefix area (known as the row selector), hold down the button, and drag down to highlight names of all fields in the compound primary key. Your screen should resemble the one shown in Figure B-5.

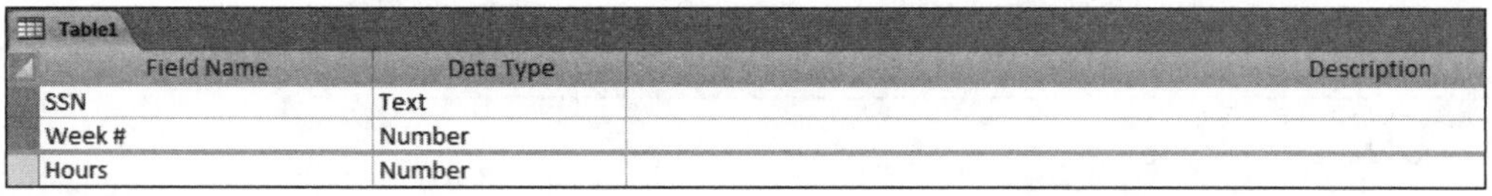

FIGURE B-5   Selecting fields for the compound primary key for the HOURS WORKED table

Now click the Key icon. Your screen should resemble the one shown in Figure B-6.

FIGURE B-6   The compound primary key for the HOURS WORKED table

That completes the creation of the compound primary key and the table definition. Use Office Button—Save to save the table as HOURS WORKED.

Adding Records to a Table

At this point, you have set up the skeletons of three tables. The tables have no data records yet. If you were to print the tables, all you would see would be column headings (the field names). The most direct way to enter data into a table is to double-click the table's name in the Navigation Pane along the left-hand side of the screen and type the data directly into the cells. (*Note:* Access 2007 uses a Navigation Pane to display and open the database objects. The Navigation Pane is located on the left-hand side of the Access window.)

AT THE KEYBOARD

On the Database window's Home tab, double-click the table Employee. Your data entry screen should resemble the one shown in Figure B-7.

FIGURE B-7   The data entry screen for the EMPLOYEE table

The EMPLOYEE table has many fields, some of which may be off the screen to the right. Scroll to see obscured fields. (Scrolling happens automatically as data is entered.) Figure B-7 shows all of the fields on the screen.

Type in your data one field value at a time. Note that the first row is empty when you begin. Each time you finish a value, press Enter; the cursor will move to the next cell. After data has been entered in the last cell in a row, the cursor moves to the first cell of the next row *and* Access automatically saves the record. (Thus, there is no need to perform the Office Button—Save step after entering data into a table.)

When entering data in your table, note that dates (for example, in the Date Hired field) should be entered as follows: 6/15/07. Access automatically expands the entry to the proper format in output.

Also note that Yes/No variables are clicked (checked) for Yes; otherwise, the box is left blank for No. You can change the box from Yes to No by clicking it, as if you were using a toggle switch.

If you make errors in data entry, click the cell, backspace over the error, and type the correction.

Enter the data shown in Figure B-8 into the EMPLOYEE table.

Last Name	First Name	SSN	Street Addre	City	State	Zip	Date Hired	US Citizen
Howard	Jane	114-11-2333	28 Sally Dr	Glasgow	DE	19702	8/1/2007	☑
Smith	John	123-45-6789	30 Elm St	Newark	DE	19711	6/1/1996	☑
Smith	Albert	148-90-1234	44 Duce St	Odessa	DE	19722	7/15/1987	☑
Jones	Sue	222-82-1122	18 Spruce St	Newark	DE	19716	7/15/2004	☐
Ruth	Billy	714-60-1927	1 Tater Dr	Baltimore	MD	20111	8/15/1999	☐
Add	Your	Data	Here	Elkton	MD	21921		☑
								☐

FIGURE B-8 Data for the EMPLOYEE table

Note that the sixth record is *your* data record. Assume that you live in Elkton, Maryland, were hired on today's date (enter the date), and are a U.S. citizen. Make up a fictitious social security number. (For purposes of this tutorial, this sixth record, as you'll discover as you continue reading, has been created using the name of one of this text's author's and the SSN 099-11-3344.)

After adding records to the EMPLOYEE table, open the WAGE DATA table and enter the data shown in Figure B-9.

SSN	Wage Rate	Salaried
114-11-2333	$10.00	☐
123-45-6789		☑
148-90-1234	$12.00	☐
222-82-1122		☑
714-60-1927		☑
Your SSN	$8.00	☐
		☐

FIGURE B-9 Data for the WAGE DATA table

In this table, you are again asked to create a new entry. For this record, enter your own SSN. Also assume that you earn $8 an hour and are not salaried. (Note that when an employee's Salaried box is not checked (i.e., Salaried = No), the implication is that the employee is paid by the hour. Because those employees who are salaried do not get paid by the hour, their hourly rate is shown as 0.00.)

Once you have finished creating the WAGE DATA table, open the HOURS WORKED table and enter the data shown in Figure B-10.

Hours Worked

SSN	Week #	Hours
114-11-2333	1	40
114-11-2333	2	50
123-45-6789	1	40
123-45-6789	2	40
148-90-1234	1	38
148-90-1234	2	40
222-82-1122	1	40
222-82-1122	2	40
714-60-1927	1	40
714-60-1927	2	40
Your SSN	1	60
Your SSN	2	55
*		

FIGURE B-10 Data for the HOURS WORKED table

Notice that salaried employees are always given 40 hours. Nonsalaried employees (including you) might work any number of hours. For your record, enter your fictitious SSN, 60 hours worked for Week 1, and 55 hours worked for Week 2.

CREATING QUERIES

Because you know how to create basic queries, this section teaches you the kinds of advanced queries you will create in the Case Studies.

Using Calculated Fields in Queries

A **calculated field** is an output field made up of *other* field values. A calculated field is *not* a field in a table; it is created in the query generator. The calculated field does not become part of the table—it is just part of query output. The best way to understand this process is to work through an example.

AT THE KEYBOARD

Suppose you want to see the SSNs and wage rates of hourly workers and you want to see what the wage rates would be if all employees were given a 10 percent raise. To view that information, show the SSN, the current wage rate, and the higher rate (which should be titled New Rate in the output). Figure B-11 shows how to set up the query.

To set up this query, you need to select hourly workers by using the Salaried field with the Criteria = No. Note in Figure B-11 that the Show box for that field is not checked, so the Salaried field values will not show in the query output.

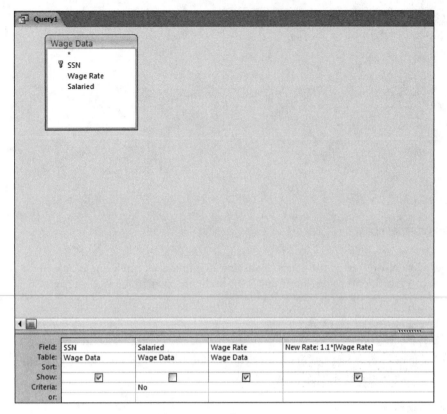

FIGURE B-11   Query setup for the calculated field

Note the expression for the calculated field, which you can see in the rightmost field cell:

New Rate: 1.1 * [Wage Rate]

The term *New Rate:* merely specifies the desired output heading. (Don't forget the colon.) The 1.1 * [Wage Rate] multiplies the old wage rate by 110 percent, which results in the 10 percent raise.

In the expression, the field name Wage Rate must be enclosed in square brackets. That is a rule. *Any time an Access expression refers to a field name, the expression must be enclosed in square brackets.*

If you run this query, your output should resemble that shown in Figure B-12.

SSN	Wage Rate	New Rate
114-11-2333	$10.00	11
148-90-1234	$12.00	13.2
099-11-3344	$8.00	8.8

FIGURE B-12   Output for a query with calculated field

Notice that the calculated field output is not shown in Currency format; it's shown as a Double—a number with digits after the decimal point. To convert the output to Currency format, you should select the output column by clicking the line above the calculated field expression, thus activating the column, which subsequently darkens. Your data entry screen should resemble the one shown in Figure B-13.

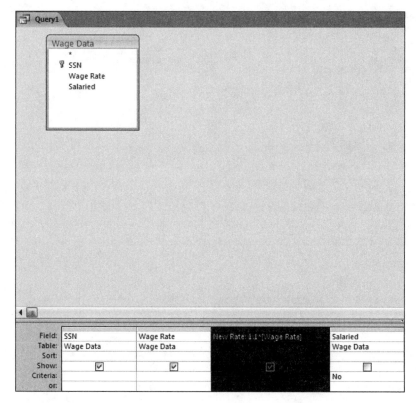

FIGURE B-13 Activating a calculated field in query design

Then on the Design tab header, click **Property Sheet** in the Show/Hide group. A Field Properties window, such as the one shown on the right in Figure B-14, will appear.

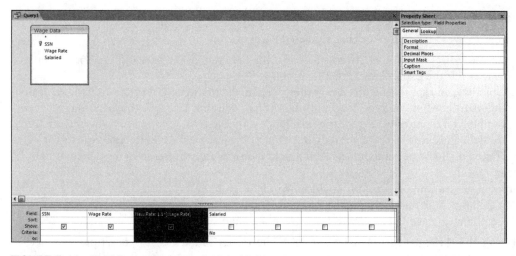

FIGURE B-14 Field Properties of a calculated field

Click **Format** and choose Currency, as shown in Figure B-15. Then click the upper right *X* to close the window.

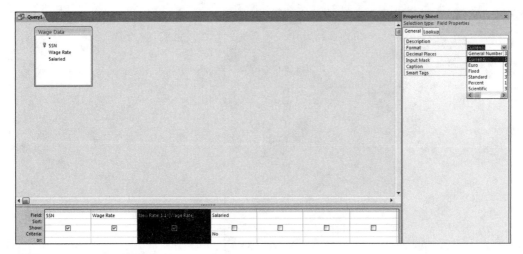

FIGURE B-15   Currency format of a calculated field

Now when you run the query, the output should resemble that shown in Figure B-16.

SSN	Wage Rate	New Rate
114-11-2333	$10.00	$11.00
148-90-1234	$12.00	$13.20
099-11-3344	$8.00	$8.80

FIGURE B-16   Query output with formatted calculated field

Next, you'll look at how to avoid errors when making calculated fields.

Avoiding Errors When Making Calculated Fields

Follow these guidelines to avoid making errors in calculated fields:

- Don't put the expression in the *Criteria* cell as if the field definition were a filter. You are making a field, so put the expression in the *Field* cell.
- Spell, capitalize, and space a field's name *exactly* as you did in the table definition. If the table definition differs from what you type, Access thinks you're defining a new field by that name. Access then prompts you to enter values for the new field, which it calls a Parameter Query field. That is easy to debug because of the tag Parameter Query. If Access asks you to enter values for a Parameter, you almost certainly misspelled a field name in an expression in a calculated field or criterion.

  Example: Here are some errors you might make for Wage Rate:
      Misspelling: (Wag Rate)
      Case change: (wage Rate / WAGE RATE)Spacing change: (WageRate / Wage Rate)

- Don't use parentheses or curly braces instead of the square brackets. Also, don't put parentheses inside square brackets. You *are*, however, allowed to use parentheses outside the square brackets, in the normal algebraic manner.

  Example: Suppose you want to multiply Hours times Wage Rate to get a field called Wages Owed. This is the correct expression:
      Wages Owed: [Wage Rate] * [Hours]
  The following also would be correct:
      Wages Owed: ([Wage Rate] * [Hours])
  But it would *not* be correct to omit the inside brackets, which is a common error:
      Wages Owed: [Wage Rate * Hours]

"Relating" Two (or More) Tables by the Join Operation

Often the data you need for a query is in more than one table. To complete the query, you must join the tables, linking the common fields known as a *join*. One rule of thumb is that joins are made on fields that have common *values*, and those fields often can be key fields. The names of the join fields are irrelevant; also, the names of the tables (or fields) to be joined may be the same, but that is not a requirement for an effective join.

Make a join by bringing in (Adding) the tables needed. Next, decide which fields you will join. Then click one field name and hold down the left mouse button while you drag the cursor over to the other field's name in its window. Release the button. Access puts in a line, signifying the join. (*Note:* If a relationship between two tables has been formed elsewhere, Access will put in the line automatically, in which case you do not have to do the click-and-drag operation. Often Access puts in join lines without the user forming relationships.)

You can join more than two tables. The common fields *need not* be the same in all tables; that is, you can daisy-chain them together.

A common join error is to Add a table to the query and then fail to link it to another table. In that case, you will have a table floating in the top part of the QBE (query by example) screen. When you run the query, your output will show the same records over and over. That error is unmistakable because there is *so much* redundant output. The rules are (1) add only the tables you need and (2) link all tables.

Next, you'll work through an example of a query needing a join.

AT THE KEYBOARD

Suppose you want to see the last names, SSNs, wage rates, salary status, and citizenship only for U.S. citizens and hourly workers. Because the data is spread across two tables, Employee and Wage Data, you should add both tables and pull down the five fields you need. Then you should add the Criteria. Set up your work to resemble that shown in Figure B-17. Make sure the tables are joined on the common field, SSN.

Here is a quick review of the Criteria you will need to set up this join: If you want data for employees who are U.S. citizens *and* who are hourly workers, the Criteria expressions go in the *same* Criteria row. If you want data for employees who are U.S. citizens *or* who are hourly workers, one of the expressions goes in the second Criteria row (the one that has the or: notation).

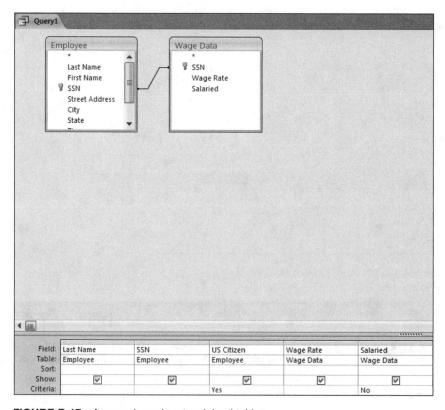

FIGURE B-17 A query based on two joined tables

Now run the query. The output should resemble that shown in Figure B-18, with the exception of the name Brady.

Last Name	SSN	US Citizen	Wage Rate	Salaried
Howard	114-11-2333	☑	$10.00	☐
Smith	148-90-1234	☑	$12.00	☐
Brady	099-11-3344	☑	$8.00	☐
*		☐		☐

FIGURE B-18   Output of a query based on two joined tables

As there is no need to print the query output or save it, you should go back to the Design View and close the query. Another practice query follows.

AT THE KEYBOARD

Suppose you want to see the wages owed to hourly employees for Week 2. To do that, you should show the last name, the SSN, the salaried status, the week #, and the wages owed. Wages will have to be a calculated field ([Wage Rate] * [Hours]). The criteria are No for Salaried and 2 for the Week # (*Note:* This means another "And" query is required). Your query should be set up like the query displayed in Figure B-19.

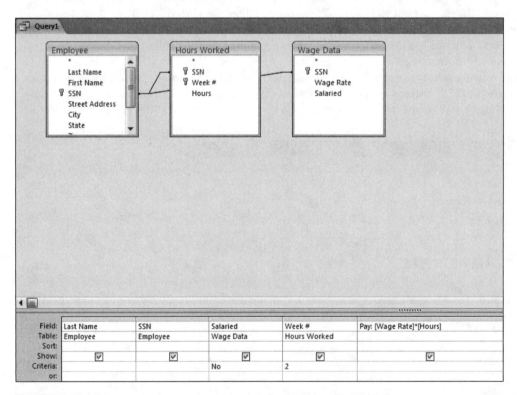

FIGURE B-19   Query setup for wages owed to hourly employees for Week 2

NOTE

Note that in the query in Figure B-19, the calculated field column was widened so you can see the whole expression. To widen a column, click the column boundary line and drag to the right.

Run the query. The output should be similar to that shown in Figure B-20 (if you formatted your calculated field to Currency).

Query1

Last Name	SSN	Salaried	Week #	Pay
Howard	114-11-2333	☐	2	$500.00
Smith	148-90-1234	☐	2	$480.00
Brady	099-11-3344	☐	2	$440.00
*		☐		

FIGURE B-20 Query output for wages owed to hourly employees for Week 2

Notice that it was not necessary to pull down the Wage Rate and Hours fields to make the query work. As there is no need to print the query output or save it, you should go back to the Design View and close the query.

Summarizing Data from Multiple Records (Totals Queries)

You may want data that summarizes values from a field for several records (or possibly all records) in a table. For example, you might want to know the average hours that all employees worked in a week or perhaps the total (sum of) all of the hours worked. Furthermore, you might want data grouped (stratified) in some way. For example, you might want to know the average hours worked, grouped by all U.S. citizens versus all non-U.S. citizens. Access calls that kind of query a **Totals query**. Those operations include the following:

Sum	The total of a given field's values
Count	A count of the number of instances in a field—that is, the number of records. (In the current example, to get the number of employees, you'd count the number of SSN numbers.)
Average	The average of a given field's values
Min	The minimum of a given field's valuest
Var	The variance of a given field's values
StDev	The standard deviation of a given field's values
Where	The field has criteria for the query output

AT THE KEYBOARD

Suppose you want to know how many employees are represented in the example database. The first step is to bring the Employee table into the QBE screen. Do that now. Since you will need to count the number of SSNs, which is a Totals query operation, you must bring down the SSN field.

To tell Access you want a Totals query, click the little Totals icon in the Design tab in the Show/Hide group.

That opens up a new row in the lower part of the QBE screen, called the Total row. At this point, the screen resembles that shown in Figure B-21.

FIGURE B-21   Totals query setup

Note that the Total cell contains the words *Group By*. Until you specify a statistical operation, Access assumes a field will be used for grouping (stratifying) data.

To count the number of SSNs, click next to Group By, which reveals a little arrow. Click the arrow to reveal a drop-down menu, as shown in Figure B-22.

FIGURE B-22   Choices for statistical operation in a Totals query

Select the Count operator. (With this menu, you may need to scroll to see the operator you want.) Your screen should now resemble that shown in Figure B-23.

FIGURE B-23   Count in a Totals query

Run the query. Your output should resemble that shown in Figure B-24.

FIGURE B-24   Output of Count in a Totals query

Notice that Access made a pseudo-heading "CountOfSSN." To do that, Access spliced together the statistical operation (Count), the word *Of*, and the name of the field (SSN). What if you wanted a phrase such as *Count of Employees* as a heading? In the Design View, you'd change the query to resemble the one shown in Figure B-25.

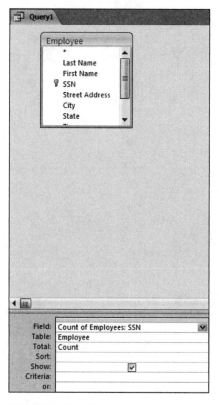

FIGURE B-25 Heading change in a Totals query

Now when you run the query, the output should resemble that shown in Figure B-26.

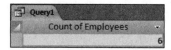

FIGURE B-26 Output of heading change in a Totals query

As there is no need to print the query output or save it, you should go back to the Design View and close the query.

Here is another example of a Totals query. Suppose you want to know the average wage rate of employees, grouped by whether the employees are salaried.

Figure B-27 shows how your query should be set up.

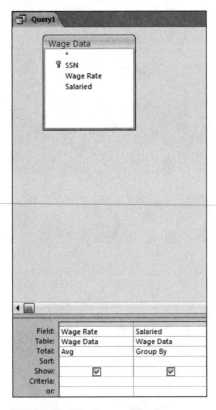

Field:	Wage Rate	Salaried
Table:	Wage Data	Wage Data
Total:	Avg	Group By
Sort:		
Show:	☑	☑
Criteria:		
or:		

FIGURE B-27   Query setup for average wage rate of employees

When you run the query, your output should resemble that shown in Figure B-28.

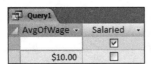

FIGURE B-28   Output of query for average wage rate of employees

Recall the convention that salaried workers are assigned zero dollars an hour. Suppose you want to eliminate the output line for zero dollars an hour because only hourly rate workers matter for that query. The query setup is shown in Figure B-29.

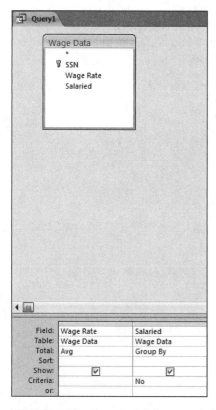

Field:	Wage Rate	Salaried
Table:	Wage Data	Wage Data
Total:	Avg	Group By
Sort:		
Show:	☑	☑
Criteria:		No
or:		

FIGURE B-29 Query setup for nonsalaried workers only

Now when you run the query, you'll get output for nonsalaried employees only, as shown in Figure B-30.

FIGURE B-30 Query output for nonsalaried workers only

Thus, it's possible to use a Criteria in a Totals query, just as you would with a "regular" query. As there is no need to print the query output or save it, you should go back to the Design View and close the query.

AT THE KEYBOARD

Assume you want to see two pieces of information for hourly workers: (1) the average wage rate—call it Average Rate in the output and (2) 110 percent of the average rate—call it the Increased Rate. To do that, you can make a calculated field in a Totals query.

You already know how to do certain things for this query. The revised heading for the average rate will be Average Rate (type *Average Rate: Wage Rate* in the Field cell). Note that you want the average of this field. Also, the grouping would be by the Salaried field. (To get hourly workers only, enter *Criteria: No.*)

The most difficult part of this query is to construct the expression for the calculated field. Conceptually, it is as follows:

Increased Rate: 1.1 * [The current average, however that is denoted]

The question is how to represent [The current average]. You cannot use Wage Rate, because that heading denotes the wages before they are averaged. Surprisingly, you can use the new heading (Average Rate) to denote the averaged amount as follows:

Increased Rate: 1.1 * [Average Rate]

Although it may seem counterintuitive, *you can treat* Average Rate *as if it were an actual field name*. Note, however, that if you use a calculated field such as Average Rate in another calculated field, as shown in Figure B-31, you must show that original calculated field in the query output. If you don't, the query will ask you to *enter parameter value*, which is incorrect. Use the setup shown in Figure B-31.

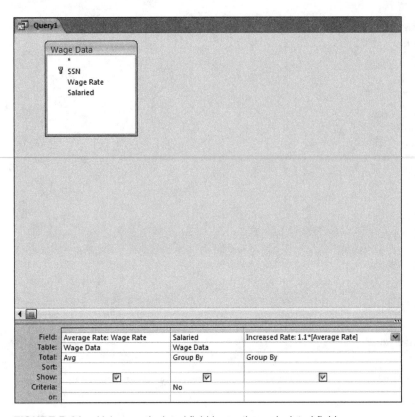

FIGURE B-31   Using a calculated field in another calculated field

Note: If you ran the query now shown in Figure B-31, you'd get some sort of error message. That is because there is no *statistical* operator applied to the calculated field's Total cell; instead, the words *Group By* appear there. To correct that, you must change the Group By operator to Expression. You may have to scroll down to get to Expression in the list.

Figure B-32 shows how the screen looks before the query is run.

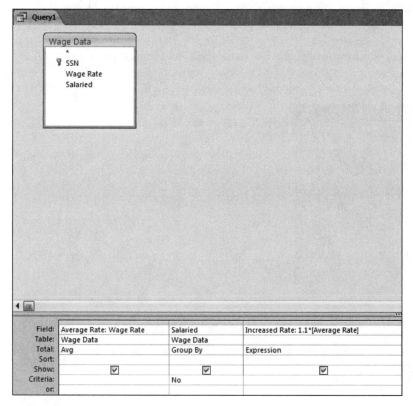

FIGURE B-32   An Expression in a Totals query

Figure B-33 shows the output of the query.

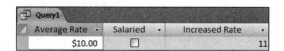

FIGURE B-33   Output of an Expression in a Totals query

As there is no need to print the query output or save it, you should go back to the Design View and close the query.

Using the Date() Function in Queries

Access has two date function features that are important for you to know. A description of each follows.

1. The following built-in function gives you today's date:

   Date()
   You can use that function in a query criteria or in a calculated field. The function "returns" the day on which the query is run; that is, it puts the value into the place where the date() function appears in an expression.

2. *Date arithmetic* lets you subtract one date from another to obtain the difference—in terms of number of days—between two calendar dates. For example, suppose you create the following expression:

   10/9/2007 – 10/4/2007
   Access would evaluate that as the integer 5 (9 less 4 is 5).

Here is another example of how date arithmetic works. Suppose you want to give each employee a bonus equaling a dollar for each day the employee has worked for you. You'd need to calculate the number of days between the employee's date of hire and the day the query is run, then multiply that number by $1.

You would find the number of elapsed days by using the following equation:

Date() – [Date Hired]

Also suppose that for each employee, you want to see the last name, SSN, and bonus amount. You'd set up the query as shown in Figure B-34.

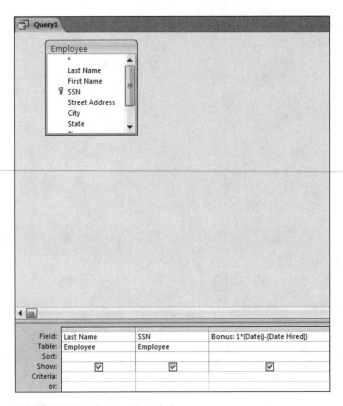

FIGURE B-34   Date arithmetic in a query

Assume you had set the format of the Bonus field to Currency. The output will be similar to that shown in Figure B-35. (Your Bonus data will be different because you are working on a date that is different from the date this tutorial was written.)

Last Name	SSN	Bonus
Brady	099-11-3344	$0.00
Howard	114-11-2333	$179.00
Smith	123-45-6789	$3,892.00
Smith	148-90-1234	$7,136.00
Jones	222-82-1122	$926.00
Ruth	714-60-1927	$2,722.00

FIGURE B-35   Output of query with date arithmetic

Using Time Arithmetic in Queries

Access also allows you to subtract the values of time fields to get an elapsed time. Assume your database has a JOB ASSIGNMENTS table that shows the times that nonsalaried employees were at work during a day. The definition is shown in Figure B-36.

Table1	
Field Name	**Data Type**
ᵠ SSN	Text
ClockIn	Date/Time
ClockOut	Date/Time
DateWorked	Date/Time

FIGURE B-36   Date/Time data definition in the JOB ASSIGNMENTS table

Assume the Date field is formatted for Long Date and the ClockIn and ClockOut fields are formatted for Medium Time. Also assume that for a particular day, nonsalaried workers were scheduled as shown in Figure B-37.

Table1	
Field Name	**Data Type**
ᵠ SSN	Text
ClockIn	Date/Time
ClockOut	Date/Time
DateWorked	Date/Time

FIGURE B-37   Display of date and time in a table

You want a query that will show the elapsed time that your employees were on the premises for the day. When you add the tables, your screen may show the links differently. Click and drag the JOB ASSIGNMENTS, EMPLOYEE, and WAGE DATA table icons to look like those in Figure B-38.

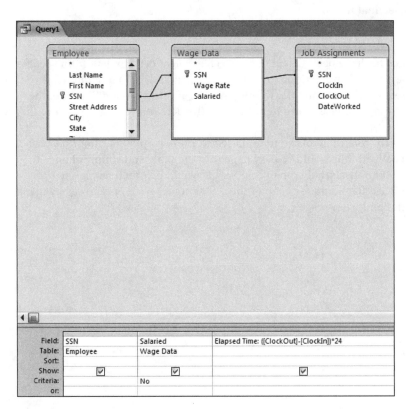

FIGURE B-38   Query setup for time arithmetic

Figure B-39 shows the output.

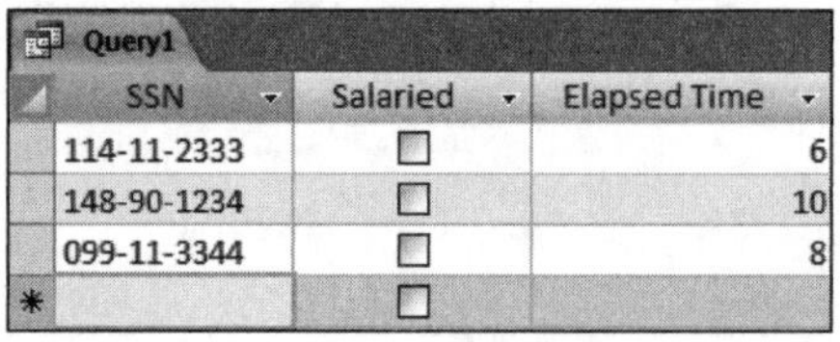

Query1

SSN	Salaried	Elapsed Time
114-11-2333	☐	6
148-90-1234	☐	10
099-11-3344	☐	8
*	☐	

FIGURE B-39 Query output for time arithmetic

The output looks correct. For example, employee 099-11-3344 was at work from 8:30 a.m. to 4:30 p.m., which is eight hours. But how does the odd expression that follows yield the correct answers?

([ClockOut] – [ClockIn]) * 24

Why wouldn't the following expression work?

[ClockOut] – [ClockIn]

Here is the answer: In Access, subtracting one time from the other yields the *decimal* portion of a 24-hour day. Returning to the example, you can see that employee 099-11-3344 worked eight hours, which is one-third of a day, so the time arithmetic function yields .3333. That is why you must multiply by 24—to convert from decimals to an hourly basis. Hence, for employee 099-11-3344, the expression performs the following calculation: 1/3 x 24 = 8.

Note that parentheses are needed to force Access to do the subtraction *first*, before the multiplication. Without parentheses, multiplication takes precedence over subtraction. For example, consider the following expression:

[ClockOut] – [ClockIn] * 24

In that case, ClockIn would be multiplied by 24, the resulting value would be subtracted from ClockOut, and the output would be a nonsensical decimal number.

Deleting and Updating Queries

The queries presented in this tutorial thus far have been Select queries. They select certain data from specific tables based on a given criterion. You also can create queries to update the original data in a database. Businesses do that often, and they do it in real time. For example, when you order an item from a Web site, the company's database is updated to reflect the purchase of the item through the deletion of that item from the company's inventory.

Consider an example. Suppose you want to give all of the nonsalaried workers a $0.50 per hour pay raise. With the three nonsalaried workers you have now, it would be easy to go into the table and simply change the Wage Rate data. But assume you have 3,000 nonsalaried employees. Now it would be much faster—not to mention more accurate—to change the Wage Rate data for each of the 3,000 nonsalaried employees by using an Update query that adds $0.50 to each employee's wage rate.

Now you will change each of the nonsalaried employees' pay via an Update query. Figure B-40 shows how to set up the query.

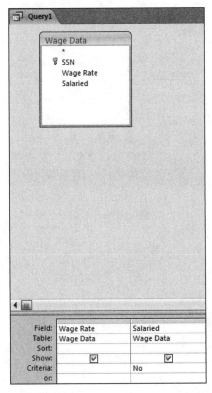

FIGURE B-40   Query setup for an Update query

So far, this query is just a Select query. Click the Update button in the Query Type group, as shown in Figure B-41.

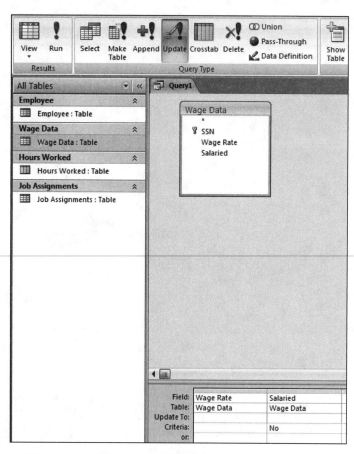

FIGURE B-41   Selecting a query type

Notice that you now have another line on the QBE grid called *Update to:*. That is where you specify the change or update to the data. Notice that you are going to update only the nonsalaried workers by using a filter under the Salaried field. Update the Wage Rate data to Wage Rate plus $0.50 as shown in Figure B-42. (Note that the update involves the use of brackets [ ], as in a calculated field.)

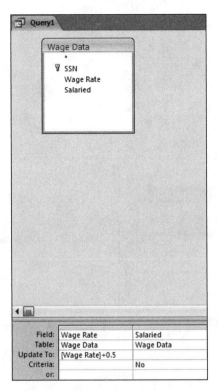

FIGURE B-42   Updating the wage rate for nonsalaried workers

Now run the query by clicking the Run button in the Results group. If you are unable to run the query because it's blocked by Disabled Mode, you need to choose Database Tools tab, Message Bar in the Show/Hide group. Click the options button and choose enable this content, then OK. When you successfully run the query, you will get a warning message as shown in Figure B-43.

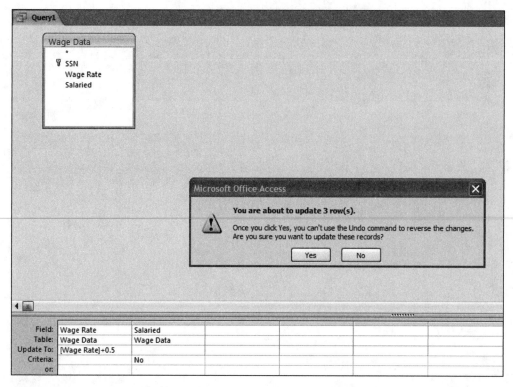

FIGURE B-43 Update Query warning

Once you click **Yes**, the records will be updated. Check those updated records now by viewing the Wage Data table. Each salaried wage rate should be increased by $0.50. Note that in this example, you are simply adding $0.50 to each salaried wage rate. You could add or subtract data from another table as well. If you do that, remember to put the field name in square brackets.

Another kind of query is the Delete query. Delete queries work like Update queries. Assume your company has been taken over by the state of Delaware. The state has a policy of employing only Delaware residents. Thus, you must delete (or fire) all employees who are not exclusively Delaware residents. To do that, you need to create a Select query. Using the EMPLOYEE table, choose the Delete icon from Query Type group; then bring down the State field and filter only those records not in Delaware (DE). Do not perform the operation, but note that if you did, the setup would look like that shown in Figure B-44.

FIGURE B-44 Deleting all employees who are not Delaware residents

Using Parameter Queries

Another kind of query, which is actually a type of Select query, is a **Parameter query**. Here is an example: Suppose your company has 5,000 employees and you want to query the database to find the same kind of information again and again, but about different employees each time. For example, you might want to query the database to find out how many hours a particular employee has worked. To do that, you could run a query that you created and stored previously, but run it only for a particular employee.

Create a Select query with the format shown in Figure B-45.

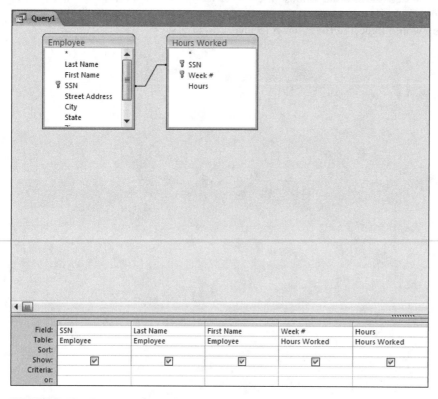

FIGURE B-45 Design of a Parameter query begins as a Select query

In the Criteria line of the QBE grid for the field SSN, type what is shown in Figure B-46.

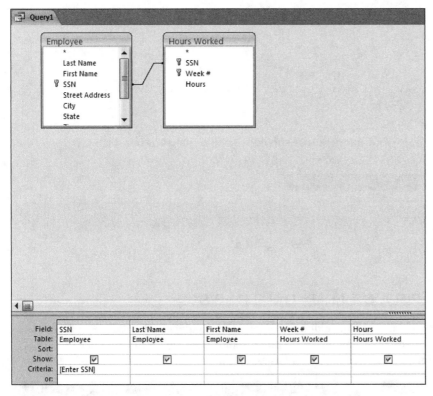

FIGURE B-46   Design of a Parameter query, continued

Note that the Criteria line involves the use of square brackets, as you would expect to see in a calcu-lated field.

Now run the query. You will be prompted for the specific employee's SSN, as shown in Figure B-47.

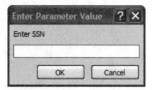

FIGURE B-47 Enter Parameter Value dialog box

Type in your own (fictitious) SSN. Your query output should resemble that shown in Figure B-48.

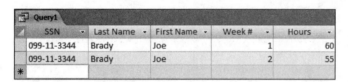

FIGURE B-48 Output of a Parameter query

MAKING SEVEN PRACTICE QUERIES

This portion of the tutorial is designed to give you additional practice in making queries. Before making these queries, you must create the specified tables and enter the records shown in the Creating Tables section of this tutorial. The output shown for the practice queries is based on those inputs.

AT THE KEYBOARD

For each query that follows, you are given a problem statement and a "scratch area." You also are shown what the query output should look like. Follow this procedure: Set up a query in Access. Run the query. When you are satisfied with the results, save the query and continue with the next query. *Note:* You will be working with the EMPLOYEE, HOURS WORKED, and WAGE DATA tables.

1. Create a query that shows the SSN, last name, state, and date hired for those employees who are currently living in Delaware *and* were hired after 12/31/99. Sort (ascending) by SSN. (A recap of sorting procedures: Click the Sort cell of the field. Choose Ascending or Descending.) Before creating your query, use the table shown in Figure B-49 to work out your QBE grid on paper.

FIGURE B-49 CBE grid template

Field					
Table					
Sort					
Show					
Criteria					
Or:					

Your output should resemble that shown in Figure B-50.

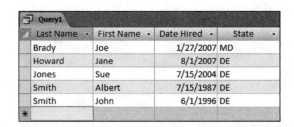

Query1			
SSN	Last Name	State	Date Hired
114-11-2333	Howard	DE	8/1/2007
222-82-1122	Jones	DE	7/15/2004
*			

FIGURE B-50   Number 1 query output

2. Create a query that shows the last name, first name, date hired, and state for those employees who are currently living in Delaware *or* were hired after 12/31/99. The primary sort (ascending) is on last name, and the secondary sort (ascending) is on first name. (Sorting recap: The Primary Sort field must be to the left of the Secondary Sort field in the query setup.) Before creating your query, use the table shown in Figure B-51 to work out your QBE grid on paper.

FIGURE B-51   QBE grid template

Field					
Table					
Sort					
Show					
Criteria					
Or:					

If your name was Joe Brady, your output would look like that shown in Figure B-52.

Query1			
Last Name	First Name	Date Hired	State
Brady	Joe	1/27/2007	MD
Howard	Jane	8/1/2007	DE
Jones	Sue	7/15/2004	DE
Smith	Albert	7/15/1987	DE
Smith	John	6/1/1996	DE
*			

FIGURE B-52   Number 2 query output

3. Create a query that shows the sum of hours that U.S. citizens worked and the same sum for non-U.S. citizens (that is, group on citizenship). The heading for total hours worked should be Total Hours Worked. Before creating your query, use the table shown in Figure B-53 to work out your QBE grid on paper.

FIGURE B-53   QBE grid template

Field					
Table					
Total					
Sort					
Show					
Criteria					
Or:					

Your output should resemble that shown in Figure B-54.

Total Hours Worked	US Citizen
363	☑
160	☐

FIGURE B-54 Number 3 query output

4. Create a query that shows the wages owed to hourly workers for Week 1. The heading for the wages owed should be Total Owed. The output headings should be Last Name, SSN, Week #, and Total Owed. Before creating your query, use the table shown in Figure B-55 to work out your QBE grid on paper.

FIGURE B-55 QBE grid template

Field					
Table					
Sort					
Show					
Criteria					
Or:					

If your name was Joe Brady, your output would look like that shown in Figure B-56.

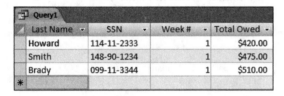

Last Name	SSN	Week #	Total Owed
Howard	114-11-2333	1	$420.00
Smith	148-90-1234	1	$475.00
Brady	099-11-3344	1	$510.00

FIGURE B-56 Number 4 query output

5. Create a query that shows the last name, SSN, hours worked, and overtime amount owed for hourly employees who earned overtime during Week 2. Overtime is paid at 1.5 times the normal hourly rate for all of the hours worked over 40. Note that the amount shown in the query should be just the overtime portion of the wages paid. Also, this is not a Totals query—amounts should be shown for individual workers. Before creating your query, use the table shown in Figure B-57 to work out your QBE grid on paper.

FIGURE B-57 QBE grid template

Field					
Table					
Sort					
Show					
Criteria					
Or:					

If your name was Joe Brady, your output would look like that shown in Figure B-58.

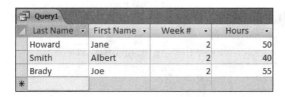

FIGURE B-58 Number 5 query output

6. Create a Parameter query that shows the hours employees have worked. Have the Parameter query prompt for the week number. The output headings should be Last Name, First Name, Week #, and Hours. Do this only for the nonsalaried workers. Before creating your query, use the table shown in Figure B-59 to work out your QBE grid on paper.

FIGURE B-59 QBE grid template

Field					
Table					
Sort					
Show					
Criteria					
Or:					

Run the query using 2 when prompted for the Week #. Your output should look like that shown in Figure B-60.

Last Name	First Name	Week #	Hours
Howard	Jane	2	50
Smith	Albert	2	40
Brady	Joe	2	55

FIGURE B-60 Number 6 query output

7. Create an update query that gives certain workers a merit raise. First, you must create an additional table as shown in Figure B-61.

FIGURE B-61 MERIT RAISES table

Now make a query that adds the Merit Raise to the current Wage Rate for those employees who will receive a raise. When you run the query, you should be prompted with *You are about to update two rows.* Check the original WAGE DATA table to confirm the update. Before creating your query, use the table shown in Figure B-62 to work out your QBE grid on paper.

FIGURE B-62 QBE grid template

Field					
Table					
Update to					
Criteria					
Or:					

CREATING REPORTS

Database packages let you make attractive management reports from a table's records or from a query's output. If you are making a report from a table, the Access report generator looks up the data in the table and puts it into report format. If you are making a report from a query's output, Access runs the query in the background (you do not control it or see it happen) and then puts the output in report format.

There are three ways to make a report. One is to handcraft the report from scratch in the Design View. Because that is a tedious process, it is not shown in this tutorial. The second way is to use the Report Wizard, during which Access leads you through a menu-driven construction. That method is shown in this tutorial. The third way is to start in the Wizard and then use the Design View to tailor what the Wizard produces. That method also is shown in this tutorial.

Creating a Grouped Report

This tutorial assumes that you already know how to use the Wizard to make a basic ungrouped report. This section of the tutorial teaches you how to make a grouped report. (If you don't know how to make an ungrouped report, you can learn by following the first example in the upcoming section.)

AT THE KEYBOARD

Suppose you want to make a report out of the HOURS WORKED table. Choose the Create tab, Report Wizard in the Reports group. Select the HOURS WORKED table from the drop-down menu as the report basis. Select all of the fields (using the **>>** button), as shown in Figure B-63.

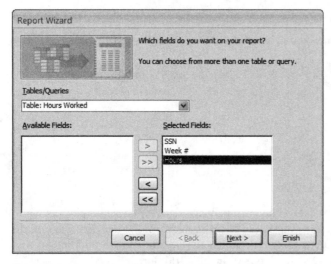

FIGURE B-63 Field selection step in the Report Wizard

Click **Next**. Then tell Access you want to group on Week # by double-clicking that field name. This grouping step is shown in Figure B-64.

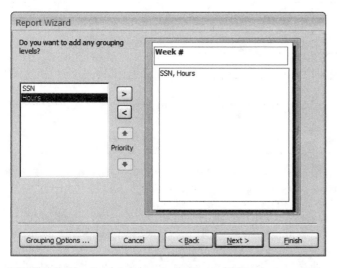

FIGURE B-64   Grouping step in the Report Wizard

Click **Next**. You'll see a screen, similar to the one in Figure B-65, for Sorting and Summary Options.

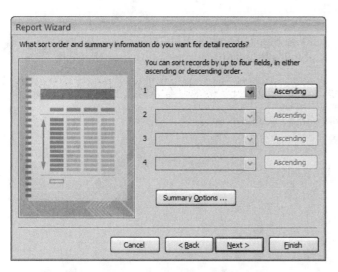

FIGURE B-65   Sorting and Summary Options step in the Report Wizard

Because you chose a grouping field, Access will let you decide whether you want to see group subtotals and/or report grand totals. If you choose that option, all numeric fields could be added. In this example, group subtotals are for total hours in each week. Assume you *do* want the total of hours by week. Click **Summary Options**. You'll get a screen similar to the one in Figure B-66.

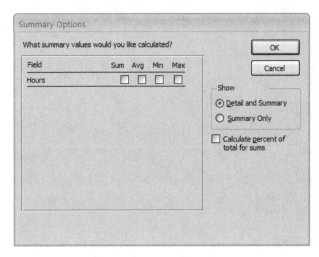

FIGURE B-66 Summary Options in the Report Wizard

Next, follow these steps:

1. Click the Sum box for Hours (to sum the hours in the group).
2. Click **Detail and Summary**. (Detail equates with "group"; Summary, with "grand total for the report.")
3. Click **OK**. This takes you back to the Sorting screen, where you can choose an ordering within the group if desired. (In this case, you don't choose one.)
4. Click **Next** to continue.
5. In the Layout screen (not shown here), choose Stepped and Portrait.
6. Make sure the "Adjust the field width so all fields fit on a page" check box is unchecked.
7. Click **Next**.
8. In the Style screen (not shown), accept Office.
9. Click **Next**.
10. Provide a title—Hours Worked by Week would be appropriate.
11. Select the Preview the report button to view the report.
12. Click **Finish**.

Your report will look like that shown in Figure B-67.

Hours Worked by Week

Week #	SSN	Hours
1		
	099-11-3344	60
	714-60-1927	40
	222-82-1122	40
	148-90-1234	38
	123-45-6789	40
	114-11-2333	40

Summary for 'Week #' = 1 (6 detail records)

Sum 258

FIGURE B-67 Hours Worked by Week report

Notice that data is grouped by weeks, with Week 1 on top, then a subtotal for that week. Week 2 data is next, then a grand total (which you can scroll down to see). The subtotal is labeled "Sum," which is not very descriptive. That can be changed later in the Design View. Also, there is the apparently useless "Summary for

Week . . ." line, which can be deleted later in the Design View as well. Click the *x* in the upper right-hand corner of your report window corresponding to "Close Hours Worked by Week." Note that the Navigation Pane on the left side of your screen contains the report shown in Figure B-68.

FIGURE B-68   Navigation Pane with report

To edit the report in the Design View, right-click the report title and choose Design view. You will see a complex (and intimidating) screen, similar to the one shown in Figure B-69.

FIGURE B-69   Report design screen

The organization of the screen is hierarchical. At the top is the Report level. The next level down (within a report) is the Page level. The next levels down (within a page) are for any data groupings you have specified.

If you told Access to make group (summary) totals, your report would have a Report Header area and end with a Grand Total in the Report Footer. The report header is usually just the title you specified.

A page also has a header, which is usually just the names of the fields you told Access to put in the report (here, Week #, SSN, and Hours fields). Sometimes the page number is put in by default.

Groupings of data are more complex. There is a header for the group. In this case, the *value* of the Week # will be the header; so the values shown will be 1 and 2. Those headers indicate that there is a group of data for the first week, then a group of data for the second week. Within each data grouping is the other "detail" that you've requested. In this case, there will be data for each SSN and the related hours.

Each Week # gets a footer, which is a labeled sum. Recall that you asked the footer to be shown (i.e., Detail and Summary were requested). The Week # Footer is composed of three elements:

1. The line that starts =Summary for . . .
2. The Sum label
3. The adjacent expression =Sum(Hours)

The line beneath the Week # Footer will be printed unless you eliminate it. Similarly, the word *Sum* will be printed as the subtotal label unless you eliminate it. The =Sum(Hours) is an expression that tells Access to add the quantity *for the header in question* and to put that number into the report as the subtotal. (In this example, that number would be the sum of hours by Week #.)

Each report also gets a footer—the grand total (in this case, of hours) for the report.

If you look closely, each of the detail items appears to be doubly inserted in the design. For example, in Figure B-69, you will see the notation for SSN twice, once in the Page Header band and then again in the Detail band. Hours are treated similarly.

Those items will not, however, be printed twice, because each detail item in the report is an object. In Access, an object is denoted by both a label and its value. In other words, there is a representation of the name of the object, which is the boldfaced name itself (in this example, *SSN* in the Page Header band), and there is a representation in less bold type of the value of the object (corresponding to *SSN* in the Detail band).

Sometimes the Report Wizard is arbitrary about where it puts labels and data. If you do not like where the Wizard puts data, you can move the objects containing the data around in the Design View. Also, you can click and drag within the band or across bands. Often a box will be too small to allow the full numerical values for that object to show. When that happens, select the box and click one of the sides to stretch it. Doing so will allow full values to show. At other times, an object's box will be very long. When that happens, you can click the box, resize it, then drag right or left in its panel to reposition the output.

Suppose you do *not* want the Summary for . . . line to appear in the report. Also suppose you would like different subtotal and grand total labels. The Summary for . . . line is an object you can activate by clicking. Do that now. "Handles" (little squares) should appear around its edges, as shown in Figure B-70.

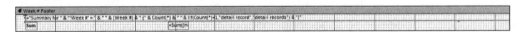

FIGURE B-70 Selecting an object in the Report Design View

Press the Delete key to get rid of the selected object.
To change the subtotal heading, click the Sum object, as shown in Figure B-71.

FIGURE B-71 Selecting the Sum object in the Report Design View

Click again. That gives you an insertion point from which you can type, as shown in Figure B-72.

FIGURE B-72 Typing an object in the Report Design View

Change the label to Sum of Hours for Week and press Enter or click somewhere else in the report to deactivate the object. Your screen should resemble that shown in Figure B-73.

FIGURE B-73 Changing a label in the Report Design View

Tutorial B

You can change the Grand Total in the same way.

In the Design tab, Click the View button on the Views group. You should see a report similar to that in Figure B-74 (top part is shown).

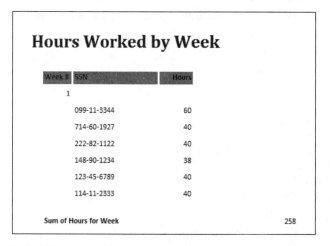

FIGURE B-74   Hours Worked by Week report

Notice that the data are grouped by week number (data for Week 1 is shown) and subtotaled for that week. The report also has a grand total at the bottom.

Moving Fields in the Design View

If you want to group on more than one field in the Report Wizard, the report will have an odd "staircase" look (or display repeated data) or it will have both features. Next, you will learn how to overcome that effect in the Design View.

Suppose you make a query showing an employee's last name, first name, street address, zip code, week #, and hours worked. Then you make a report from that query, grouping on last name and first name only. That is shown in Figure B-75.

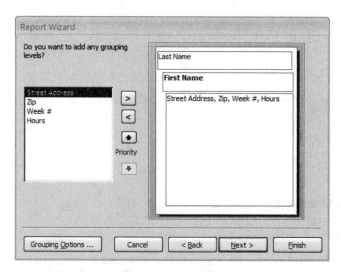

FIGURE B-75   Grouping in the Report Wizard

Then follow these steps:

1. Click **Next**.
2. Do not Sum anything in Summary Options. Click **Next**.
3. Select Stepped.

4. Select Landscape.
5. Click the check mark by *Adjust the field width so all fields fit on a page*. Click **Next**.
6. Select Office. Click **Next**.
7. Type a title (Hours Worked by Employees). Click **Finish**.

As you preview the report, you will notice repeating data. In the report displayed in Figure B-76, notice that the Zip data and Street Address data are repeating and that both are shown below the First Name data, which is below the Last Name data—hence, the staircase effect. (The fields Week # and Hours are shown subordinate to Last Name and First Name, as desired.)

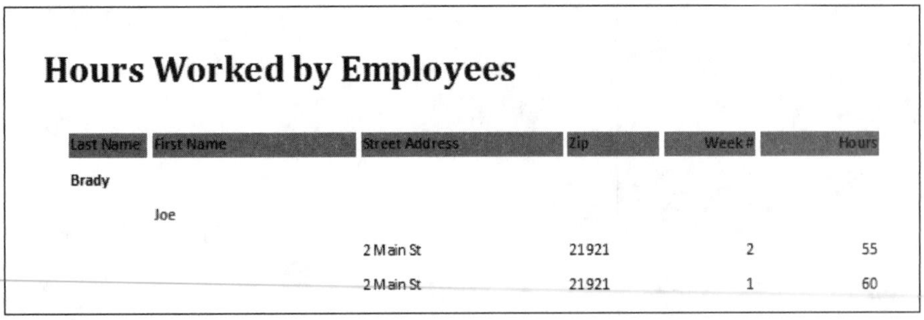

Hours Worked by Employees

Last Name	First Name	Street Address	Zip	Week #	Hours
Brady					
	Joe				
		2 Main St	21921	2	55
		2 Main St	21921	1	60

FIGURE B-76 Hours Worked by Employees grouped report

Suppose you want the last name, first name, street address, and zip all to appear on the same line. The way to do that is to take the report into the Design View for editing. From the Navigation Pane on the left side of your screen, right-click the Hours Worked by Employees report and choose Design View. At this point, the headers look like those shown in Figure B-77.

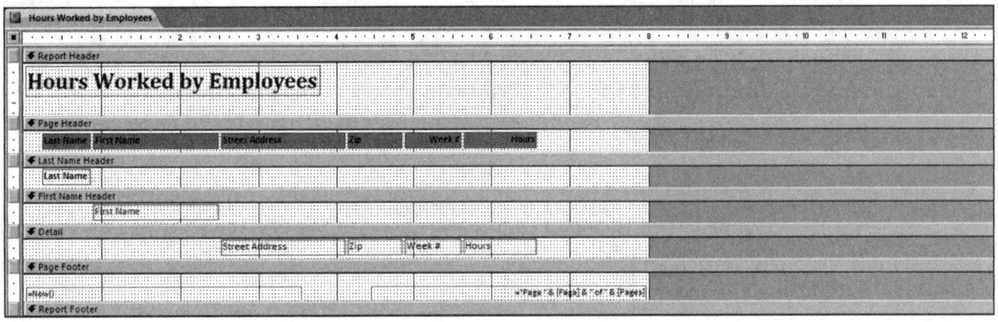

FIGURE B-77 Hours Worked by Employees report displayed in Design View

Your goal is to get the First Name, Street Address, and Zip fields into the Last Name header band (*not* into the Page Header band) so they will print on the same line. The first step is to click the First Name object in the First Name Header band, as shown in Figure B-78.

FIGURE B-78 Selecting First Name object in the First Name header

Right-click, cut, place your cursor in the Last Name header, and paste into the Last Name Header, as shown in Figure B-79. (*Note:* When you paste the First Name object, it will overlay the Last Name object. To correct that, simply move the First Name object to the right side of the Last Name object.)

FIGURE B-79 Moving the First Name object to the Last Name header

Now cut the Street Address object and paste it into the Last Name header as shown in Figure B-80.

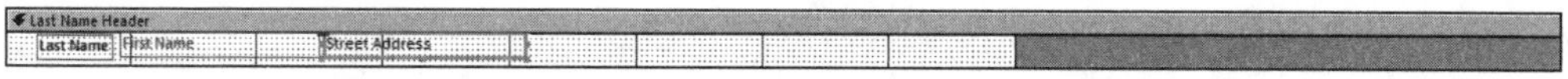

FIGURE B-80   Moving the Street Address object to the Last Name header

Do the same for the Zip object. Now correct the spacing and the Page Header by choosing the Layout View from the View group on the Design tab. Click **Hours** and note that the entire column is selected (your heading and data). Drag the column to the right side of the report. At this point, your screen should look like that shown in Figure B-81.

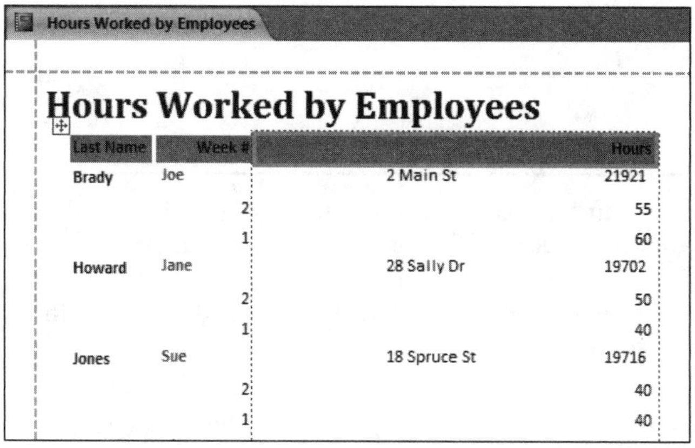

FIGURE B-81   Adjusting report layout

Drag the left side of the Hours column to the right; the Week # heading and data will follow your cursor to the right. Your report should now resemble the one shown in Figure B-82.

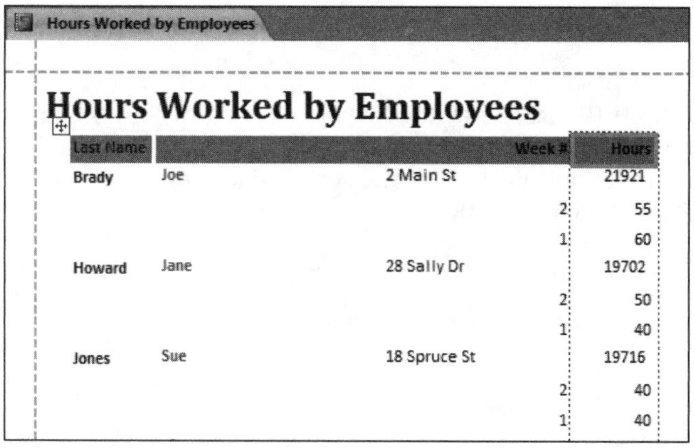

FIGURE B-82   Adjusting report layout, continued

Finally, adjust the headings by dragging the left edge of the Week # to the right and adding more descriptions such as First Name, Address, and Zip to the Last Name header. This final adjustment will result in a report similar to that shown in Figure B-83. (*Note:* This is the Report View.)

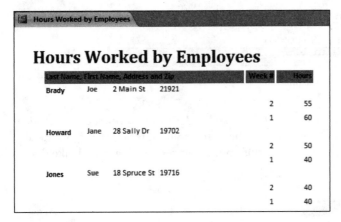

FIGURE B-83   Hours Worked by Employees report

IMPORTING DATA

Text or spreadsheet data is easily imported into Access. In business, it is often necessary to import data because companies use disparate systems. Assume your healthcare coverage data is on the human resources manager's computer in an Excel spreadsheet.

Open the software application Microsoft Excel. Create a spreadsheet in Excel using the data shown in Figure B-84.

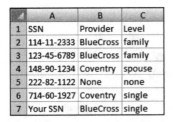

	A	B	C
1	SSN	Provider	Level
2	114-11-2333	BlueCross	family
3	123-45-6789	BlueCross	family
4	148-90-1234	Coventry	spouse
5	222-82-1122	None	none
6	714-60-1927	Coventry	single
7	Your SSN	BlueCross	single

FIGURE B-84   Excel data

Save the file; then close it. Now you can easily import that spreadsheet data into a new table in Access. With your **Employee** database open, go to the External Data tab, Import group and click **Excel**. Browse to find the Excel file you just created and make sure the first radio button (Import the source data into a new table in the current database.) is selected, as shown in Figure B-85. Click **OK**.

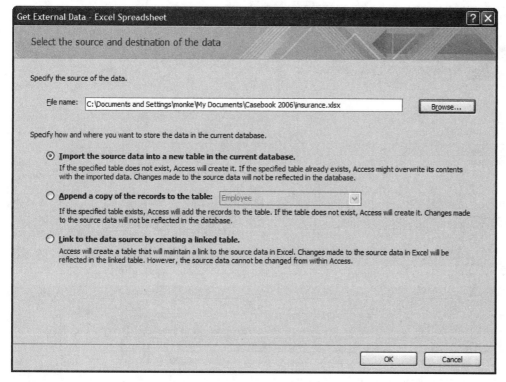

FIGURE B-85 Importing Excel data into a new table

Choose the correct worksheet. Assuming you have just one worksheet in your Excel file, your next screen should look like that shown in Figure B-86.

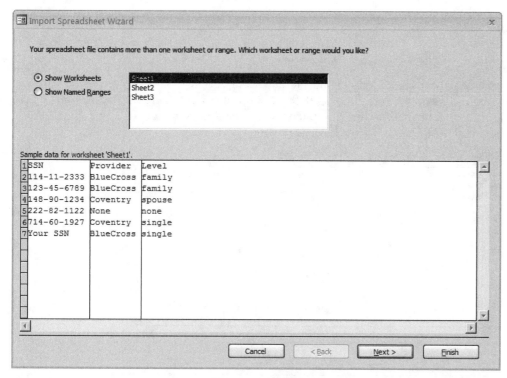

FIGURE B-86 First screen in the Import Spreadsheet Wizard

Choose Next and make sure you select the box that says First Row Contains Column Headings, as shown in Figure B-87.

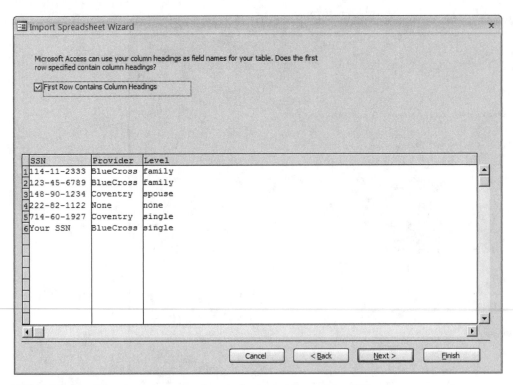

FIGURE B-87 Choosing column headings in the Import Spreadsheet Wizard

Choose Next. Accept the default for each field you are importing on this screen. Each field is assigned a text data type, which, in this case, is correct for this table. Your screen should look like that shown in Figure B-88.

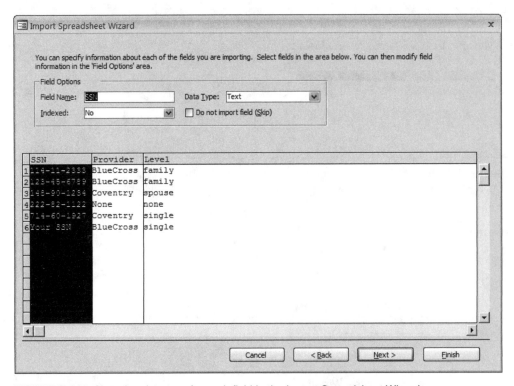

FIGURE B-88 Choosing data type for each field in the Import Spreadsheet Wizard

Choose Next. In the next screen of the Wizard, you'll be prompted to create an index—that is, define a primary key. Since you are going to be storing your data in a new table, choose your own primary key (SSN), as shown in Figure B-89.

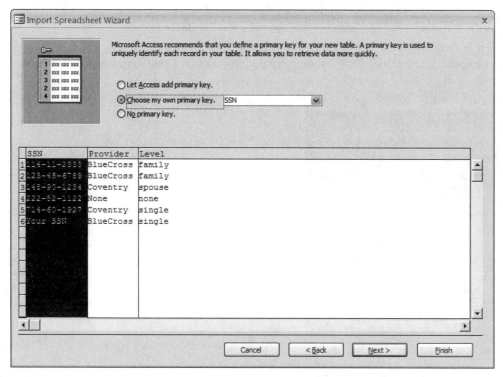

FIGURE B-89   Choosing a primary key field in the Import Spreadsheet Wizard

Continue through the Wizard, giving your table an appropriate name. After importing the table, take a look at it and its design. (Highlight the Table option and use the Design button.) Note the width of each field (very large). Adjust the field properties as needed.

MAKING FORMS

Forms simplify the process of adding new records to a table. The Form Wizard is easy to use and can be applied to a single table or to multiple tables.

When you base a form on one table, you simply identify that table when you are in the Form Wizard setup. The form will then have all of the fields from that table and only those fields. When data is entered into the form, a complete new record is automatically added to the table. Forms with two tables are discussed next.

Making Forms with Subforms

You also can make a form that contains a subform, which can be useful when you need to create a form that is based on two (or more) tables. Return to the example Employee database to see how forms and subforms would be particularly handy for viewing all of the hours that each employee worked each week. Suppose you want to show all of the fields from the EMPLOYEE table; you also want to show the hours each employee worked (in other words, include all fields from the HOURS WORKED table as well).

Creating the Form and Subform

To create the form and subform, first create a simple one-table form using the Form Wizard on the EMPLOYEE table. Follow these steps:

1. In the Create tab, Forms group, choose More Forms, Form Wizard.
2. Make sure the table Employee is selected under the drop-down menu of Tables/Queries.
3. Select all Available Fields by clicking the right double-arrow button.

4. Select Next.
5. Select Columnar layout.
6. Select Next.
7. Select Office Style.
8. Select Next.
9. When asked "What title do you want for your form?" type *Employee Hours*.
10. Select Finish.

After the form is complete, in the Design tab, Views group, choose Design View so your screen looks like the one shown in Figure B-90.

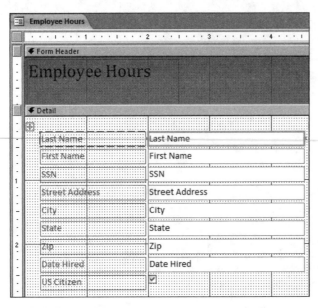

FIGURE B-90   The Employee Hours form

In the Design tab, Controls group, locate the Subform/Subreport button, which is located right above the letter *C* in *Controls*, as shown in Figure B-91.

FIGURE B-91   The Controls group

Click the Subform/Subreport button (sixth row, button on right) and using your cursor, drag a small section next to the State, Zip, Date Hired, and US Citizen fields in your form design. As you lift your cursor, the Subform Wizard will appear, as shown in Figure B-92. (*Note:* You may need to stretch the area of the form before placing the subform.)

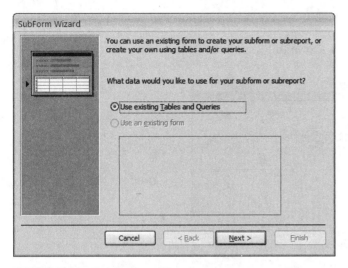

FIGURE B-92   The Subform Wizard

Follow these steps to create data in the subform:

1. Select the button Use existing Tables and Queries.
2. Select Next.
3. Under Tables/Queries, choose the HOURS WORKED table and bring all fields into the Selected Fields box by clicking the right double-arrow button.
4. Select Next.
5. Select the Choose from a list radio button.
6. Select Next.
7. Use the default subform name.
8. Select Finish.

Now you need to adjust the design so the data of all of the fields are visible. Go to the Datasheet View and click through the various records to see how the subform data changes. Your final form should resemble the one shown in Figure B-93.

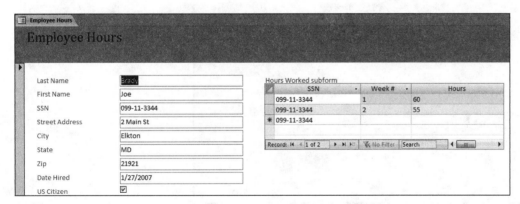

FIGURE B-93   The Employee Hours form with the Hours Worked subform

CREATING A CUSTOM NAVIGATION PANE

If you want someone who knows nothing about Access to be able to run your database, you can create a customized area, known as a Navigation Pane, to simplify the work. A Navigation Pane provides a simple, user-friendly interface that has groupings of objects that the user can click to perform certain tasks. For example, you can design a custom Navigation Pane with two groupings: one for the Forms and one for the Reports. Your finished product will show the objects within each grouping along the left side of the screen. The user can open each object by simply clicking on it.

To design that custom Navigation Pane, right-click the top of the Navigation Pane and choose Navigation Options.

The Navigation Options dialog box will open, as shown in Figure B-94.

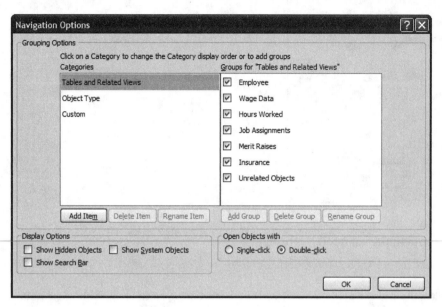

FIGURE B-94   The Navigation Options dialog box

Click the Add Item button, which appears below the Categories column, as shown in Figure B-95.

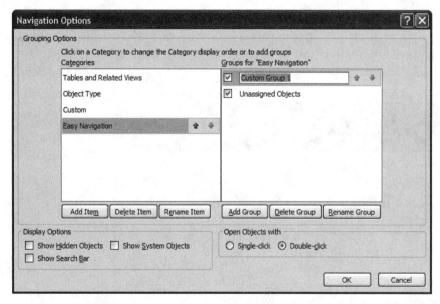

FIGURE B-95   Add Item in the Navigation Options dialog box

Change the Custom Group 1 name to Easy Navigation and press the Enter key.

With the Easy Navigation still selected, click the Add Group button, which appears below the right-hand column, Groups for "Easy Navigation." Add a group for Forms and a group for Reports. Leave the Unassigned Objects group as is. Your screen should look like the one shown in Figure B-96.

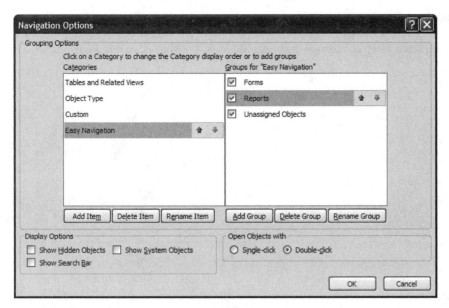

FIGURE B-96   Adding groups to Easy Navigation

Choose OK. Your screen should look like the one shown in Figure B-97.

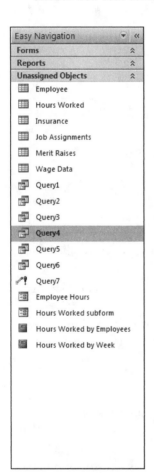

FIGURE B-97   Easy Navigation without objects

Now you need to add objects to your new groups. Choose Easy Navigation from the top of the Navigation Pane.

Drag each form in the Unassigned Objects group to the Forms group and drag each report to the Reports group. When you are finished, you can hide the Unassigned Objects group by clicking the double arrow in the Unassigned Objects bar. Your completed Easy Navigation Pane should look like that shown in Figure B-98.

FIGURE B-98 Easy Navigation completed

TROUBLESHOOTING COMMON PROBLEMS

Access beginners (and veterans) sometimes create databases that have problems. Some common problems are described below, along with their causes and corrections.

1. *"I saved my database file, but I can't find it on my computer or external secondary storage medium! Where is it?"*

    You saved your file to some fixed disk or some place other than My Documents. Use the Search option of the Windows Start button. Search for all files ending in .accdb (search for *.accdb). If you did save the file, it is on the hard drive (C:\) or on some network drive. (Your site assistant can tell you the drive designators.)

2. *"What is a 'duplicate key field value'? I'm trying to enter records into my Sales table. The first record was for a sale of product X to customer 101, and I was able to enter that one. But when I try to enter a second sale for customer #101, Access tells me I already have a record with that key field value. Am I allowed to enter only one sale per customer?"*

    Your primary key field needs work. You may need a compound primary key—customer number and some other field(s). In this case, customer number, product number, and date of sale might provide a unique combination of values; or you might consider using an invoice number field as a key.

3. *"My query says, 'Enter Parameter Value' when I run it. What is that?"*

   This symptom, 99 times out of 100, indicates that you have an expression in a Criteria or a calculated field in which *you have misspelled a field name*. Access is very fussy about spelling. For example, Access is case-sensitive. The program is also "space-sensitive," which means when you put a space in a field name when you define the table, you must put a space in the field name when you reference it in a query expression. Fix the typo in the query expression.

4. *"I'm getting a fantastic number of rows in my query output—many times more than I need. Most of the rows are duplicates!"*

   This symptom is usually caused by a failure to link all of the tables you brought into the top half of the query generator. The solution is to use the manual click-and-drag method to link the common fields between tables. (Spelling of the field names is irrelevant because the link fields need not be spelled the same.)

5. *"For the most part, my query output is what I expected; but I am getting one or two duplicate rows or not enough rows."*

   You may have linked too many fields between tables. Usually, only a single link is needed between two tables. It's unnecessary to link each common field in all combinations of tables; usually, it's enough to link the primary keys. A layperson's explanation for why overlinking causes problems is that excess linking causes Access to "overthink" the problem and repeat itself in its answer. On the other hand, you might be using too many tables in the query design. For example, you brought in a table, linked it on a common field with some other table, but then did not use the table—in other words, you brought down none of its fields and/or you used none of its fields in query expressions. In this case, if you were to get rid of the table, the query would still work. Try doing the following to see whether the few duplicate rows disappear: Click the unneeded table's header in the top of the QBE area and press the Delete key.

6. *"I expected six rows in my query output, but I only got five. What happened to the other one?"*

   Usually, this indicates a data entry error in your tables. When you link the proper tables and fields to make the query, remember that the linking operation joins records from the tables *on common values* (*equal* values in the two tables). For example, if a primary key in one table has the value "123", the primary key or the linking field in the other table should be the same to allow linking. Note that the text string "123" is not the same as the text string "123 "—the space in the second string is considered a character too. Access does not see unequal values as an error. Instead, Access moves on to consider the rest of the records in the table for linking. Solution: Look at the values entered into the linked fields in each table and fix any data entry errors.

7. *"I linked fields correctly in a query, but I'm getting the empty set in the output. All I get are the field name headings!"*

   You probably have zero common (equal) values in the linked fields. For example, suppose you are linking on Part Number (which you declared as text). In one field, you have part numbers "001", "002", and "003"; and in the other table, you have part numbers "0001", "0002", and "0003". Your tables have no common values, which means no records are selected for output. You'll have to change the values in one of the tables.

8. *"I'm trying to count the number of today's sales orders. A Totals query is called for. Sales are denoted by an invoice number, and I made that a text field in the table design. However, when I ask the Totals query to 'Sum' the number of invoice numbers, Access tells me I cannot add them up! What is the problem?"*

   Text variables are words! You cannot add words, but you can count them. Use the Count Totals operator (not the Sum operator) to count the number of sales, each being denoted by an invoice number.

9. *"I'm doing time arithmetic in a calculated field expression. I subtracted the Time In from the Time Out and got a decimal number! I expected eight hours, and I got the number .33333. Why?"*

[Time Out] – [Time In] yields the decimal percentage of a 24-hour day. In your case, eight hours is one-third of a day. You must complete the expression by multiplying by 24: ([Time Out] – [Time In]) * 24. Don't forget the parentheses.

10. *"I formatted a calculated field for Currency in the query generator, and the values did show as currency in the query output; however, the report based on the query output does not show the dollar sign in its output. What happened?"*

Go into the report Design View. A box in one of the panels represents the calculated field's value. Click the box and drag to widen it. That should give Access enough room to show the dollar sign as well as the number in the output.

11. *"I told the Report Wizard to fit all of my output to one page. It does print to one page, but some of the data is missing. What happened?"*

Access fits the output all on one page by *leaving data out*. If you can stand to see the output on more than one page, click off the Fit to a Page option in the Wizard. One way to tighten output is to go into the Design View and remove space from each of the boxes representing output values and labels. Access usually provides more space than needed.

12. *"I grouped on three fields in the Report Wizard, and the Wizard prints the output in a staircase fashion. I want the grouping fields to be on one line. How can I do that?"*

Make adjustments in the Design View and the Layout View. See the Reports section of this tutorial for instruction on how to make those adjustments.

13. *"When I create an Update query, Access tells me that zero rows are updating or more rows are updating than I want. What is wrong?"*

If your Update query is not set up correctly (for example, if the tables are not joined properly), Access will try not to update anything or will update all of the records. Check the query, make corrections, and run it again.

14. *"After making a Totals Query with a Sum in the Group By row and saving that query, when I go back to it, the Sum field says Expression and Sum is put in the field name box. Is that wrong?"*

Access sometimes changes that particular statistic when the query is saved. The data remains the same, and you can be assured your query is correct.

15. *"I am not able to run my Update Query, but I know it is set up correctly. What is wrong?"*

Check the Security Content of the database by clicking the Security Content button. You may need to enable certain actions.

PRELIMINARY CASE: THE COLLEGE GEEKS

Setting up a Relational Database to Create Tables, a Form, Queries, and a Report

PREVIEW

In this case, you'll create a relational database for a small company that sets up and repairs computers. First, you'll create three tables and populate them with data. Next, you'll create a form for recording visits to clients and three queries: a Select query, a Count query, and a query with a Calculated Field. Finally, you'll create a report that outlines the jobs and the money that each job brings in each month.

PREPARATION

Before attempting this case, you should:

- Have some experience using Microsoft Access.
- Complete any part of Access Tutorial B that your instructor assigns or refer to the tutorial as necessary.

BACKGROUND

Like students at any other university, students at the University of Kansas have problems with their computers. Although many college students are adept at working on computers, the majority of college students are not knowledgeable about computer security, diagnostics, repair, or setup. Therefore, at most universities, student-run companies fix fellow students' computers. The College Geeks is a small company started at the University of Kansas by three junior-year business students who do just that.

The company is very busy this fall term, helping students get rid of viruses, set up their computers, and fix hardware problems. One of the owners of the College Geeks heard that you are proficient in Microsoft Access. He hired you to finish the database project begun in the summer to help keep track of customers and jobs. The database tables are already designed for you. There are three tables in the database:

1. The PRICE LIST AND JOB DESCRIPTION table keeps track of Job ID numbers, a Description of the types of jobs that are handled, and the Price charged Per Hour of work.
2. The STUDENTS table keeps track of all customers (who are students). It lists their Student ID, Last Name, First Name, Dorm Address, Cell Phone, and E-mail.
3. The VISITS table records the Visit ID number for each visit, the Date on which the customer is visited, the type of work that is performed (Job ID), the customer's Student ID number, and the Number of Hours the visit takes.

The owners would like to see a few additional capabilities in the database:

- The ability to record each visit quickly and simply. This task can be accomplished with a form.
- The ability to have the database answer some questions. First, the owners would like to know which jobs were performed in a specific dorm. That information would be helpful when conducting future marketing campaigns since the company's reputation is spread by word of mouth. Also, students who live together in dorms are often friends who have similar computer problems because they often do similar tasks on their computers.

- The ability to track how many times each type of job is performed to further improve future marketing campaigns. Thus, the owners need a listing of jobs from most visited to least visited. In addition, as the business grows, hiring new employees would be easier if the business owners knew what specialties job applicants should have.
- The ability to track the amount of money each student customer owes.
- The ability to compile a summary report that shows all of the money brought in by each job each month.

ASSIGNMENT 1: CREATING TABLES

Use Microsoft Access to create the tables with the fields shown in Figures 1-1 through 1-3 and discussed in the Background section. Populate the database tables as shown. Add your name to the STUDENTS table with a fictitious student ID. Complete the table with your dorm address, cell phone number, and e-mail address.

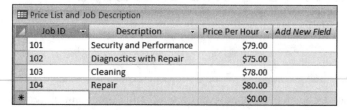

Price List and Job Description			
Job ID	Description	Price Per Hour	Add New Field
101	Security and Performance	$79.00	
102	Diagnostics with Repair	$75.00	
103	Cleaning	$78.00	
104	Repair	$80.00	
*		$0.00	

FIGURE 1-1   The PRICE LIST AND JOB DESCRIPTION table

Students						
Student ID	Last Name	First Name	Dorm Address	Cell Phone	Email	Add New Field
112312	Peters	Sally	114 Jayhawker	785-665-0980	Sally@ku.edu	
112679	Brown	Jerome	609 Jayhawker	785-453-9822	jbb@ku.edu	
554686	Marvel	Thomas	678-C Lewis	785-223-9812	TM@ku.edu	
654761	Paulson	Hunter	891-A Lewis	785-908-5411	hunter@ku.edu	
876579	Rodriquez	John	8-B Sellards	554-983-9263	RodQ@ku.edu	
899043	Chou	Patti	10-D Sellards	785-665-1432	pc90@ku.edu	
899044	Friedman	Amy	715 Jayhawker	785-665-3324	amy@ku.edu	
988976	French	Allison	115 Jayhawker	456-908-8872	Alli@ku.edu	
*						

FIGURE 1-2   The STUDENTS table

Visits					
Visit ID	Date	Job ID	Student ID	Number of Hours	Add New Field
1	10/15/2007	103	112312	3	
10	11/3/2007	101	899043	3	
2	10/15/2007	103	876579	2	
3	10/16/2007	101	876579	2	
4	10/16/2007	101	988976	2	
5	10/18/2007	102	554686	2	
6	11/1/2007	104	112312	4	
7	11/1/2007	102	654761	2	
8	11/3/2007	104	112679	1	
9	11/3/2007	104	899044	1	
*				0	

FIGURE 1-3   The VISITS table

ASSIGNMENT 2: CREATING A FORM, QUERIES, AND A REPORT

Assignment 2A: Creating a Form

Using only the VISITS table and the Form Wizard, create a form on which repairpersons can easily record each visit. Choose Columnar layout and any Style. Save the form as Visits. View one record and print only one record.

Assignment 2B: Creating a Select Query

Create a Select Query that displays information about visits to the dorm Jayhawker. Display the Visit ID (a unique number that tracks each time a geek makes a call on a customer), Date, Last Name, First Name, Dorm Address, and Cell Phone. *Hint:* When creating the criteria for the Dorm Address field, consider using wildcards. Save the query as Visits to Jayhawker. Your output should resemble that shown in Figure 1-4.

Visit ID	Date	Last Name	First Name	Dorm Address	Cell Phone
4	10/16/2007	French	Allison	115 Jayhawker	456-908-8872
1	10/15/2007	Peters	Sally	114 Jayhawker	785-665-0980
6	11/1/2007	Peters	Sally	114 Jayhawker	785-665-0980
8	11/3/2007	Brown	Jerome	609 Jayhawker	785-453-9822
9	11/3/2007	Friedman	Amy	715 Jayhawker	785-665-3324

FIGURE 1-4 Query: Visits to Jayhawker

Run the Query. Print the results.

Assignment 2C: Creating a Count Query

Create a query that counts the number of visits for each type of job. Display a listing of jobs from most visited to least visited. Your output should resemble that shown in Figure 1-5. Print the output.

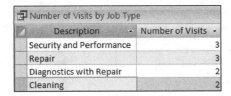

Description	Number of Visits
Security and Performance	3
Repair	3
Diagnostics with Repair	2
Cleaning	2

FIGURE 1-5 Query: Number of Visits by Job Type

Note the heading change (Number of Visits). Save the query as Number of Visits by Job Type.

Assignment 2D: Creating a Query with a Calculated Field

Create a query that calculates the money owed by each student for the computer jobs. Your output should include the Student ID, Last Name, Dorm Address, and Money Owed and should resemble Figure 1-6. Print the output.
Note the heading of the calculated field. Save the query as Money Owed by Student.

Assignment 2E: Generating a Report

Generate a report that shows the money brought in each month, including details about all of the jobs. To create the report, you will need to do the following:

- Create a query that brings all of the data together for the report. You will need the Date; Job ID, Description, and Money, which is a calculated field.
- Using the Report Wizard, base your report on the query you made in the previous bullet point.
- Group on Date. Give the report a title of Money Brought In by Date.
- In the Design View, ensure that all references to Money are in currency format.

Money Owed by Student

Student ID	Last Name	Dorm Address	Money Owed
112312	Peters	114 Jayhawker	$554.00
112679	Brown	609 Jayhawker	$80.00
554686	Marvel	678-C Lewis	$150.00
654761	Paulson	891-A Lewis	$150.00
876579	Rodriquez	8-B Sellards	$314.00
899043	Chou	10-D Sellards	$237.00
899044	Friedman	715 Jayhawker	$80.00
988976	French	115 Jayhawker	$158.00

FIGURE 1-6   Query: Money Owed by Student

- Adjust any other parts of the report to resemble that shown in Figure 1-7.
- Use Print Preview before printing to make sure the report looks correct.

Money Brought In by Date

Date by Month	Date	Job ID	Description	Money
October 2007				
	10/15/2007	103	Cleaning	$156.00
	10/15/2007	103	Cleaning	$234.00
	10/16/2007	101	Security and Performance	$158.00
	10/16/2007	101	Security and Performance	$158.00
	10/18/2007	102	Diagnostics with Repair	$150.00
Total for the Month				$856.00
November 2007				
	11/1/2007	102	Diagnostics with Repair	$150.00
	11/1/2007	104	Repair	$320.00
	11/3/2007	101	Security and Performance	$237.00
	11/3/2007	104	Repair	$80.00
	11/3/2007	104	Repair	$80.00
Total for the Month				$867.00
Grand Total				$1,723.00

FIGURE 1-7   Report: Money Brought In by Date

If you are working with a disk, make sure you close the database file before removing your disk.

DELIVERABLES

Assemble the following deliverables for your instructor:

1. Printouts of three tables
2. Form: Visits (one record printed)
3. Query 1: Visits to Jayhawker
4. Query 2: Number of Visits by Job Type
5. Query 3: Money Owed by Student
6. Report: Money Brought In by Date
7. Any other required tutorial printouts, tutorial disk, or CD

Staple all pages together. Put your name and class number at the top of the page. If required, make sure your disk or CD is labeled.

CASE **2**

THE INTERNET BOOK SWAP

Designing a Relational Database to Create Tables, a Form, Queries, and a Report

PREVIEW

In this case, you'll design a relational database for an Internet business that swaps books. After your database design is completed and correct, you will create database tables and populate them with data. Then you will produce one form, three queries, and one report. The queries will address the following: What author has written a book in a specific category? How many credits for potential book swaps does each member have? What books are available under a certain genre? Your report will list the current books available, grouped by genre.

PREPARATION

Before attempting this case, you should:

- Have some experience designing databases and using Microsoft Access.
- Complete any part of Database Design Tutorial A that your instructor assigns.
- Complete any part of Access Tutorial B that your instructor assigns or refer to the tutorial as necessary.
- Refer to Tutorial F as necessary.

BACKGROUND

You are an avid reader; but as a poor college student, you cannot afford to buy yourself books. Up to this point, you have relied on your neighborhood library to feed your passion for books. However, after moving on campus, you can no longer return books in a timely fashion to the local library. You can borrow books from the university library, but you have found that they get misplaced easily in your messy dorm room—and if they are not returned on time, your grades are withheld! Recently, you came up with a way to get hold of books with little money changing hands and no late fees—an Internet book swap.

You convince your closest friends to start this business with you. The plan is straightforward: people who want to swap books register on your Web site as "members." They send you *physical* books to swap and gain one credit per book they send in. You post the titles of these books on the Web site to show they are available for swap, thereby allowing other members to request the books. The only money involved is a yearly membership fee and nominal shipping costs. You split the work with your friends. Because you are proficient with Microsoft Access, you take control of designing and implementing your company's database. Your other friends split up the Web interface work and the shipping logistics.

You have several goals that you want to accomplish with the database. First, you need to set up database tables to track the following information:

- All of the current club members, using their member ID numbers and names.
- All of the books sent to the swap: each book's ID number, title, author, identification as a paperback or hardbound book (recorded for future warehouse planning), and category (or genre) of the subject matter. In addition, the books that are taken by another member—that is, traded permanently—need to be designated as being "out of circulation."
- Which member has sent in which book and which member has requested which book.

Until the Web site is up and running, your friend working on shipping logistics is in charge of receiving the e-mail messages requesting specific books. You want to create a form so she can record which member has sent in which book and note what books are no longer available because they have been shipped. Eventually, this form will be put on the Internet.

You and others will need to interact with the database. Queries such as the following could facilitate those interactions:

- Ultimately, members visiting your company's Web site will want to search for books and authors in specific genres by all or part of a book's title or by an author's name. Until your Web site is established, you need to create a prototype query to enable you and your employees to do this searching. The query will eventually be migrated to the Internet so customers can search your database themselves.
- You need a way to figure out how many "credits" a given member has. You can accomplish this by setting up a Select query that will do the calculation.
- You want to be able to answer an important question that members are likely to ask: What books are available by specific genre? You'll need to develop a Parameter query to do this.

Finally, your Web designer would like to post a daily report of which books are available, grouped by genre, for members and potential members to peruse.

ASSIGNMENT 1: CREATING THE DATABASE DESIGN

In this assignment, you design your database tables on paper, using a word processing program. Pay close attention to the tables' logic and structure. Do not start your Access code (Assignment 2) before getting feedback from your instructor on Assignment 1. Keep in mind that you need to look at the requirements for Assignment 2 to design your fields and tables properly. Good programming practice is to look at the required outputs before designing your database. When designing the database, observe the following guidelines:

- Determine the tables you'll need by listing on paper the name of each table and the fields that it should contain. Avoid data redundancy. Do not create a field if it can be created by a "calculated field" in a query.
- Include logical fields that answer questions such as these: Is the book a paperback? Has the book been swapped?
- Plan on transaction tables. Think about what business events correspond to each member's actions. Avoid duplicating data.
- Document your tables by using the Table facility of your word processor. Your word-processed tables should resemble the format of the table in Figure 2-1.
- Mark the appropriate key field(s). You can designate a key field by inserting an asterisk (*) next to the field name. Keep in mind that some tables need a compound primary key to uniquely identify a record within a table.
- Print the database design.

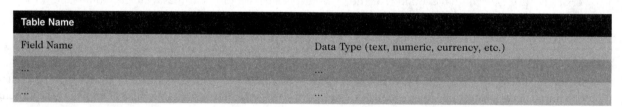

Table Name	
Field Name	Data Type (text, numeric, currency, etc.)
...	...
...	...

FIGURE 2-1 Table design

> **NOTE**
>
> Have your design approved before beginning Assignment 2; otherwise, you may need to redo Assignment 2.

ASSIGNMENT 2: CREATING THE DATABASE AND MAKING QUERIES AND A REPORT

In this assignment, you create database tables in Access and populate them with data. Then you create a form, three queries, and a report.

Assignment 2A: Creating Tables in Access

In this part of the assignment, you create your tables in Access. Use the following guidelines:

- Type records into the tables using the members' names and addresses shown in Figure 2-2. Add your name, address, and e-mail address as an additional member. Make up member IDs, telephone numbers, and e-mail addresses.
- Assume that there are many genres. Some genre suggestions are home repair, mystery, sports, and travel. Create at least two more. Search on Amazon.com to populate the table with real books and authors fitting those genres. Create at least 15 records.
- Have each member swap in at least one book. Have a few members swap in more than two books, and have a few members swap out more than two books.
- Ignore any postage fees to simplify this case.
- Appropriately limit the size of the text fields; for example, a telephone number does not need to be the default setting of 50 characters in length.
- Print all tables.

| Members | | | | | |
Last Name	First Name	Address	City	State	ZIP
Smith	John	34 Redback Rd	Austin	TX	78701
Jones	Mary	9 Elm Arch Rd	Woods Hole	MA	02543
Trainer	Lisa	9000 Rt 40, Apt 25A	San Francisco	CA	94102
Russell	Heather	20 Cheswold Blvd	Cheyenne	WY	82002
Thompson	Cynthia	31 Spectrum Dr	Portland	OR	97286
Isaacs	Irving	10010 N. Barrett Dr	River Edge	NJ	07661
Downing	Charlie	200 Mac Duff Rd	Provo	UT	84601
Tansey	Paul	625 Dawson Rd	Baton Rouge	LA	70801
Bevans	Beverly	24 Raway Rd	Livonia	MI	48150

FIGURE 2-2 Data: Member Data

Assignment 2B: Creating Forms, Queries, and a Report

For this part of the assignment, generate one form, three queries, and one report as outlined in the background of this case.

Form: Books

Create a form and subform based on two tables. The main form should be based on the table that records all of the books taken in, including whether they have been swapped out. The table for the subform should include all information for swapping out the books. If your instructor so indicates, print any one record from that form. Save the form as Books. Your form should resemble the one shown in Figure 2-3.

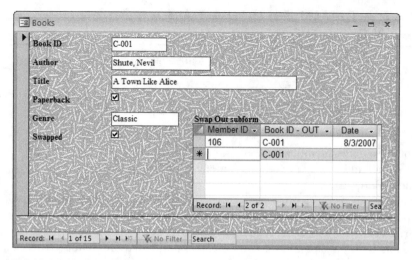

FIGURE 2-3   Form: Books

Query 1: Search on Author and Genre

Create a query called Search on Author and Genre. The output of the query should list the Author, Title, and Genre. Your query should use wildcards to search for a particular author (in case you know only part of an author's name). Choose a genre in which the author has written a book. Your output should resemble that shown in Figure 2-4, although your data will be different.

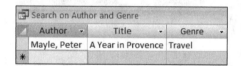

FIGURE 2-4   Query: Search on Author and Genre

Query 2: Number of Credits

Create a query called Number of Credits. This query calculates the number of credits each member has accrued. Remember, for each book a member brings in, he or she receives one credit; for each book a member obtains, he or she loses one credit. For this query, create two queries to count the number of books brought in and the number of books taken out of the book swap. Then use those queries in a new query to calculate the number of credits. Your output should list the Member ID, Last Name, and Number of Credits. Make sure you adjust the column headings so your output looks like that in Figure 2-5, although your data will be different.

Member ID	Last Name	Number of Credits
101	Smith	0
102	Jones	1
103	Trainer	0
104	Russell	1
105	Thompson	1
106	Isaacs	0
107	Downing	2
108	Tansey	1
109	Bevans	0

FIGURE 2-5   Query: Number of Credits

Query 3: Parameter by Genre

Create a query called Parameter by Genre. This query should do the following:

- Create a parameter query that will provide a list of all books available for swapping.
- Prompt the user for Genre input.
- Show the Book ID, Author, and Title and whether the book is a Paperback.

Your data may differ, but if you used the genre Travel, your output should resemble the format shown in Figure 2-6.

Book ID	Author	Title	Paperback
T-001	Mayle, Peter	A Year in Provence	☐
T-002	Automobile Association	Village Walks in Britain	☐
*			☐

FIGURE 2-6 Query: Parameter by Genre

Report

Create a report named Available Books by Genre. Your report's output should show headers for Genre, Book ID, Author, Title, and Paperback. Use the following procedure:

1. Create a query for input so that only the books available for swapping are listed.
2. Group on the field Genre.

Although your data may differ, your output should resemble that shown in Figure 2-7.

Available Books by Genre

Genre	Book ID	Author	Title	Paperback
Classic				
	C-003	Hesse, Herman	Steppenwolf	☑
Fiction				
	F-003	Adams, Douglas	Life, the Universe, and Everyt	☐
	F-002	Adams, Douglas	The Hitchhiker's Guide to the	☐
Mystery				
	M-002	Smith, Martin Cruz	Gorky Park	☐
Non-Fiction				
	N-003	Wallington, C.E.	Meterorology for Glider Pilots	☐
	N-001	Bryson, Bill	Made in America	☐
Travel				
	T-002	Automobile Association	Village Walks in Britain	☐
	T-001	Mayle, Peter	A Year in Provence	☐

FIGURE 2-7 Report: Available Books by Genre

ASSIGNMENT 3: MAKING A PRESENTATION

Create a presentation that explains the database to your colleagues. In particular, focus on what you would like to see migrated to the Web site. Include the design of your database tables and instructions for how to use the database. Your presentation should take fewer than ten minutes, including a brief question-and-answer period.

DELIVERABLES

1. Word-processed design of tables
2. Tables created in Access
3. Form: Books (one record printed)
4. Query 1: Search on Author and Genre
5. Query 2: Number of Credits
6. Query 3: Parameter by Genre
7. Report: Available Books by Genre
8. Presentation materials
9. Any other required tutorial printouts or tutorial CD

Staple all pages together. Put your name and class number at the top of the page. Make sure your disk or CD is labeled if required.

CASE **3**

THE E-COMMERCE CONVENIENCE STORE DATABASE

*Designing a Relational Database to Create a Table,
a Form, three Queries, a Report, and a Navigation Pane*

PREVIEW

In this case, you'll design a relational database for a new convenience store that physically sells items preordered from the Internet. After your database design is completed and correct, you will create database tables and populate them with data. Then you will produce a form, three queries, a report, and a navigation pane. The form will include the customer's order with order details. The queries will address the following tasks: (1) Identify the description and price of each item sold, (2) identify the most popular items among today's sales, and (3) increase the price of goods to reflect an increase in wholesale prices. Your report will summarize today's orders per customer and give the total price of the orders. You will create a custom group of objects in the navigation pane to manage the database.

PREPARATION

Before attempting this case, you should:

- Have some experience designing databases and using Microsoft Access.
- Complete any part of Database Design Tutorial A that your instructor assigns.
- Complete any part of Access Tutorial B that your instructor assigns or refer to the tutorial as necessary.
- Refer to Tutorial F as necessary.

BACKGROUND

Your Uncle Jack in Omaha, Nebraska, has decided to open a convenience store modeled after the convenience stores in Japan known as Konbini. The Japanese convenience stores are unique in that customers can order items on the Internet and then pick up and pay for the items at a designated convenience store. In addition to picking up goods at a Konbini, customers also can buy a complete cooked meal. Thus, on the way home from work, a customer can pick up a Web site order and dinner too. This is an efficient way of life for busy working people who have little time to prepare their own food—and can't wait at home for package deliveries. In Japan, a country where credit cards are not commonly used, the Konbini system also encourages e-commerce by allowing customers to pay for items ordered via the Internet with cash, check, or debit card.

Your uncle believes there is a market for a similar convenience store in Omaha. He envisions creating a store that offers high-quality prepared meals that customers can order online and then pick up and pay for on their way home from work. He knows you are proficient with databases, and he asks you to work with him this summer to create a prototype database. He'll use what you create to try to obtain bank financing for his business venture.

Jack sees the database involving these main parameters: Customers, Inventory, and Orders. Customers will be ordering their items via the Internet, so each customer's name, address, and e-mail must be recorded. Inventory also should be recorded; and to keep the database simple, Jack tells you to assume that inventory is available at all times. In other words, he has great wholesale suppliers for all of his ingredients and he can get anything he needs immediately. Of course, all orders should be recorded, and Jack thinks that most customers will be ordering many items at one time. For example, a customer might order sub sandwiches for each family member, but each sub might be different.

After you design the database, Jack wants you to create a form to be used on the Internet for ordering items. You are to create the prototype in Microsoft Access. This prototype will be very useful in Jack's presentation to the bank. For the form, you should show the main order heading, any line items the customer requests, and the quantity requested.

Jack wants to be able to query the database too. First, he would like to create queries that will display a description of the meal items and their prices. Like the order form, this, too, will eventually be migrated to the company's Web site, although it will need to be updated on a daily basis to show customers what's available for ordering. Second, Jack would like to see a list of each day's most popular items. Third, Jack would like to be able to raise prices as necessary. Occasionally, Jack's wholesale suppliers raise their prices, and Jack needs to reflect those increases in his retail prices. He would like to see how you set up a query to increase the price of a given item.

Finally, Jack needs a list of all customers who placed orders on a given day, the items they ordered, and a daily tally of how much they owe. This report would be run about 4:30 pm every day so that when rush-hour customers come in, they can pick up and pay for their items easily. You would like to make a good impression on your uncle because he is employing you for the summer. Therefore, you also propose to create a custom group of objects in the navigation pane of the database, as it would allow your uncle to easily access the information he needs.

ASSIGNMENT 1: CREATING THE DATABASE DESIGN

In this assignment, you design your database tables on paper, using a word processing program. Pay close attention to the tables' logic and structure. Do not start your Access code (Assignment 2) before getting feedback from your instructor on Assignment 1. Keep in mind, however, that you need to look at the requirements for Assignment 2 to design your fields and tables properly for Assignment 1. Good programming practice is to look at the required outputs before designing your database. When designing the database, observe the following guidelines:

- Determine the tables you'll need by listing on paper the name of each table and the fields that it should contain. Avoid data redundancy. Do not create a field if it can be created by a "calculated field" in a query.
- Create at least one transaction table. Avoid duplicating data. It is recommended that you use two tables to record transactions, one being the line item table.
- Document your tables using the Table facility of your word processor. Your word-processed tables should resemble the format of the table in Figure 3-1.
- Mark the appropriate key field(s). You can designate a key field by inserting an asterisk (*) next to the field name. Keep in mind that some tables need a compound primary key to uniquely identify a record within a table.
- Print the database design.

Table Name	
Field Name	Data Type (text, numeric, currency, etc.)
...	...
...	...

FIGURE 3-1 Table design

NOTE

Have your design approved before beginning Assignment 2; otherwise, you may need to redo Assignment 2.

ASSIGNMENT 2: CREATING THE DATABASE, A FORM, QUERIES, A REPORT, AND A NAVIGATION PANE

In this assignment, you create database tables in Access and populate them with data. Then you create three queries and a report.

Assignment 2A: Creating Tables in Access

In this part of the assignment, you create your tables in Access. Use the following guidelines:

- Type records into the tables using the customers' names and addresses shown in Figure 3-2. Add your name and address as an additional customer. Make up customer IDs, telephone numbers, and e-mail addresses.
- Assume five different types of items are for sale. Give each item a unique ID number, a description, and a price. Use an identification system that categorizes the items. For example, all salads begin with the letter S followed by a hyphen and three numerals.
- Have each customer buy at least one item. Have most customers buy more than one item. Create data for today only. You can set the Date field in the table that takes the orders to have a default of today's date. Check the Help menu for details on how to do that.
- Appropriately limit the size of the text fields; for example, a ZIP Code does not require the default setting of 255 characters in length.
- Print all tables.

Customers					
Last Name	First Name	Address	City	State	ZIP
Shepard	Patti	34 Main St	Omaha	NE	68101
Ornez	Lou	17000 Riviera Dr	Omaha	NE	68101
Black	John	5000 Woolworth Ave	Omaha	NE	68101
Douglas	Laura	802 Bauman Ave	Omaha	NE	68101
Chen	Sammy	Mason Pl	Omaha	NE	68101
Paradisi	Nancy	101 Main St	Omaha	NE	68101
Fargo	Fred	1002 N. 5th Ave	Omaha	NE	68101
Dellwin	Jacque	9 Sycamore St	Omaha	NE	68101
Patel	Sally	876 Woolworth Ave	Omaha	NE	68101
Patel	Sam	876 Woolworth Ave	Omaha	NE	68101

FIGURE 3-2 Data: Customer Data

Assignment 2B: Creating a Form, Three Queries, a Report, and a Navigation Pane

You create a form, three queries, a report, and a navigation pane as outlined in the background of this case.

Form: Orders

Create a form and subform based on the process of taking orders. Save the form as Orders. The main form should show the Order heading, specifically the Order Number, Customer ID, and Date. The subform should display the details of the order: the Order Number, Item ID, and Quantity. Print any one record from the main form. Your form should resemble that shown in Figure 3-3.

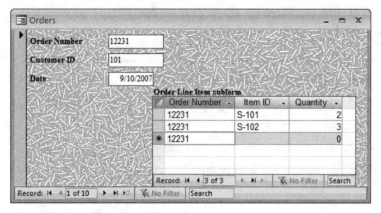

FIGURE 3-3   Form: Orders

Query 1: Description and Price of Item

Create a query called Description and Price of Item. The output of the query should list only the Name, Description, and Price of each item. Your output should resemble that shown in Figure 3-4.

Name	Description	Price
Southwest Salad	Salad with Chicken, Cheese, Chipotle Dressing	$15.50
Pad Thai	Noodles with Tofu, Peanuts, Bean Sprouts, Spicy Sa	$10.00
Hamburger	1/2 Pound Burger with L, T, Fries	$8.25
Tuna Sub	Submarine Sandwich with Tuna, Mayo, L, T	$6.00
Italian Sub	Submarine Sandwich with Ham, Cheese, Salami, L, T	$8.50
*		$0.00

FIGURE 3-4   Query: Description and Price of Item

Query 2: Popular Items

Create a query called Popular Items. Show each item's Name and Number of Orders listed in order from the highest seller to the lowest seller. Make sure you change the column headings so your output looks like that shown in Figure 3-5, although your data will be different.

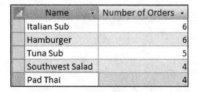

FIGURE 3-5   Query: Popular Items

Query 3: Increased Prices

Create a query called Increased Prices. This query should allow for a price increase of any item. Have the query prompt for the Item ID number of the item that is increasing in price and then prompt for the amount of the price increase. *Hint:* This is an update query with a parameter query. Assume that the price increase is for one item. Run the query and check the corresponding table for that price increase.

Report: Today's Orders

Create a report named Today's Orders. Your report's output should show headers for Customer ID, Last Name, First Name, E-mail, Item, Quantity, and Total Price. Use the following procedure:

1. Create a query for input to the report that includes calculating each customer's total order. This should be a parameter query that prompts for today's date.
2. Group the report on the Customer ID.

Case 3

3. Sum the calculated field Total Price.
4. Title the report Today's Orders.
5. Adjust the Design View so your report resembles the portion of the report shown in Figure 3-6.

Today's Orders

Customer ID	Last Name	First Name	E-mail	Item	Quantity	Total Price
101	Shepard	Patti	PBS@aol.com			
				Italian Sub	3	$25.50
				Tuna Sub	2	$12.00
Total Amount Due						*$37.50*
102	Omez	Lou	Lou123@comcast.net			
				Hamburger	1	$8.25
Total Amount Due						*$8.25*
103	Black	John	Johnny@set.net			
				Southwest Salad	1	$15.50
				Pad Thai	1	$10.00
				Hamburger	1	$8.25
Total Amount Due						*$33.75*
104	Douglas	Laura	Laura.Douglas@ibm.com			
				Pad Thai	2	$20.00

FIGURE 3-6   Report: Today's Orders

Navigation Pane: The Convenience Store

Create a custom group of objects in the navigation pane for easy access to the database. Include all of the objects in the database and categorize them as Data, Forms, Queries, and Reports. Call the custom group Easy Navigation. Its interface should resemble that shown in Figure 3-7. *Note:* The items under Data are purposely hidden here because they would reveal the database design.

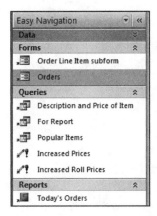

FIGURE 3-7   Navigation Pane: Easy Navigation

ASSIGNMENT 3: MAKING A PRESENTATION

Create a presentation that explains the database to your Uncle Jack. Set up the presentation so he can, in turn, use the material to make a presentation to his bank. Include the design of your database tables and describe how to use the database. Your presentation should take less than ten minutes, including a brief question-and-answer period.

DELIVERABLES

1. Word-processed design of tables
2. Tables created in Access
3. Form: Orders (one record printed)
4. Query 1: Description and Price of Item
5. Query 2: Popular Items
6. Query 3: Increased Prices (not printed)
7. Report: Today's Orders
8. Navigation Pane: Easy Navigation
9. Presentation materials
10. Any other required tutorial printouts or tutorial CD

Staple all pages together. Put your name and class number at the top of the page. Make sure your disk or CD is labeled if required.

JUST-IN-TIME LOBSTERS DATABASE

Designing a Relational Database to Create Tables, a Form, Five Queries, Two Reports, and a Navigation Pane

PREVIEW

In this case, you'll design a relational database for a seafood supply company. After your database design is completed and correct, you will create database tables and populate them with data. Then you will produce a form, five queries, two reports, and a navigation pane. The form will record all of the customers' orders. Three of the queries will display customers in a specified city, the most popular product, and the best customers, respectively; the two other queries will update the inventory to reflect purchases and orders shipped. The reports will summarize weekly sales and weekly purchases. The navigation pane will allow access to the data, form, queries, and reports.

PREPARATION

Before attempting this case, you should:

- Have some experience designing databases and using Microsoft Access.
- Complete any part of Database Design Tutorial A that your instructor assigns.
- Complete any part of Access Tutorial B that your instructor assigns or refer to the tutorial as necessary.
- Refer to Tutorial F as necessary.

BACKGROUND

Just-in-Time Lobsters (JITL) is a small, privately owned company that buys fresh lobsters and resells them to upscale restaurants in cities across the United States and Canada. JITL has specially equipped trucks that pick up the lobsters directly from suppliers. The lobsters are then transported to JITL headquarters, which is located near an air-freight hub. There, the lobsters live in an aquarium-type tank until overnight shipment to the restaurant customer is requested. This system ensures that all customers receive a fresh, consistent product wherever they are located. JITL is considering computerizing its manual files, and the company hires you for the summer to create a small prototype database to keep track of orders and answer important questions.

There are a number of parameters to keep in mind when designing the database for JITL. Using just a few suppliers, the company buys lobsters in three different sizes—medium, large, and extra large. Customers are upscale restaurants in large cities in the United States and Canada. Customers call JITL with an order, and they receive the order within 24 hours via air shipment. Shipment is included in the price of the lobsters.

When an order comes into JITL, the sales clerk assigns a unique order number and records the date and the specific customer ID number. Then under that order number, the clerk records the number of each type of lobster required. Most customers order a variety of sizes. A database form would be useful to streamline this operation.

Occasionally, the owner of JITL travels to large cities to meet with potential customers. When he travels, he likes to visit other area restaurants that are already serving JITL's lobsters. The owner would like to run a query that prompts him for the name of the city he will be visiting and then displays all of the restaurants in the city that are current customers.

Of course, any good business needs to track trends. Using database queries, the owner would like to track the most popular lobster size and JITL's best customers (in terms of dollars spent).

As orders are placed and filled, the inventory needs to be updated. That can be accomplished with an update query that's run daily after working hours. Similarly, another update query is needed to update, again on a daily basis, the inventory as purchases from suppliers are made and new lobsters are delivered to the company's tank.

To help him conduct further business analysis, the owner would like two reports created. One report should display the weekly orders from each customer of JITL, and the other report should show JITL's weekly purchases from each of its suppliers. Finally, a navigation pane should be created to simplify access to the data, form, queries, and reports.

ASSIGNMENT 1: CREATING THE DATABASE DESIGN

In this assignment, you design your database tables on paper, using a word processing program. Pay close attention to the tables' logic and structure. Do not start your Access code (Assignment 2) before getting feedback from your instructor on Assignment 1. Keep in mind that you need to look at the requirements for Assignment 2 to design your fields and tables properly. Good programming practice is to look at the required outputs before designing your database. When designing the database, observe the following guidelines:

- Determine the tables you'll need by listing on paper the name of each table and the fields that it should contain. Avoid data redundancy. Do not create a field if it can be created by a "calculated field" in a query.
- Plan on a number of transaction tables. Avoid duplicating data.
- Document your tables by using the Table facility of your word processor. Your word-processed tables should resemble the format of the table in Figure 4-1.
- Mark the appropriate key field(s). You can designate a key field by inserting an asterisk (*) next to the field name. Keep in mind that some tables need a compound primary key to uniquely identify a record within a table.
- Print the database design.

Table Name	
Field Name	Data Type (text, numeric, currency, etc.)
...	...
...	...

FIGURE 4-1 Table design

NOTE

Have your design approved before beginning Assignment 2; otherwise, you may need to redo Assignment 2.

ASSIGNMENT 2: CREATING THE DATABASE AND MAKING A FORM, QUERIES, REPORTS, AND A NAVIGATION PANE

In this assignment, you create database tables in Access and populate them with data. Then you create a form, five queries, two reports, and a navigation pane.

Assignment 2A: Creating Tables in Access

In this part of the assignment, you create your tables in Access. Use the following guidelines:

- Using the Web, research upscale restaurants in the United States and Canada. Find and create at least ten customers, typing those records into the tables.
- Assume three products are available for sale: Medium lobsters weighing at least 1.25 pounds are $30 each, large lobsters weighing at least 1.5 pounds are $42 each, and extra-large lobsters weighing at least 2 pounds are $55 each.

- Create two suppliers, one in the United States and one in Canada. Make up names, addresses, and other important information (such as telephone/fax numbers and e-mail addresses).
- Make up transaction data for one week. Assume each customer orders once during that week. Assume each supplier fills at least two orders during that week.
- Appropriately limit the size of the text fields; for example, a ZIP Code does not require the default setting of 255 characters in length.
- Print all tables.

Assignment 2B: Creating a Form, Queries, Reports, and a Navigation Pane

Create one form, five queries, two reports, and one navigation pane as outlined in the background of this case.

Form

Create an order form that includes the main information from the order, such as the Order ID, Customer ID, and Date. Include the line item details of the order in the subform. Your data will vary, but the output should resemble that shown in Figure 4-2.

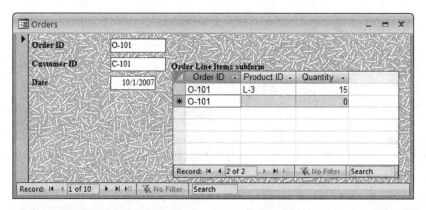

FIGURE 4-2   Form: Orders

Query 1: Customers in a Specified City

Create a query called Customers in a Specified City. This query should prompt the user to input a city name; then the query should display the Customer Name, Customer Address, and Customer Phone of the customers in that city. For example, if the city of Toronto were input, the query output might resemble that shown in Figure 4-3.

Customer Name	Customer Address	Customer Phone
Zee Grill	641 Mount Pleasant Rd	416-484-6428

FIGURE 4-3   Query 1: Customers in a Specified City

Query 2: Most Popular Product

Create a query called Most Popular Product. Display the Product Name, Product Size (pounds), and Quantity Ordered. Sort your output from the most ordered product to the least ordered product. Your data will differ, but your output should resemble that shown in Figure 4-4. Note the column heading change.

Product Name	Product Size, pounds	Quantity Ordered
Xlarge	2	60
Large	1.5	55
Medium	1.25	30

FIGURE 4-4   Query 2: Most Popular Product

Query 3: Best Customers

Create a query called Best Customers. Display the Customer Name, Customer City, Customer E-mail, and the Total Ordered, which is a calculated field. Sort your output from the highest dollar amount spent to the least dollar amount spent. Your data will differ, but the output should resemble that shown in Figure 4-5. Note the column heading change.

Customer Name	Customer City	Customer E-mail	Total Ordered
Café Pacific	Dallas	cafepac@zoom.net	$1,270.00
Zee Grill	Toronto	Zee@cbd.com	$1,050.00
Red Lobster	Wichita	RLWichita@aol.com	$825.00
Bound'Ry Restaurant	Nashville	boundry@comcast.net	$825.00
Altamar	Miami	altamar@comcast.net	$825.00
Big Fish	Dearborn	bigfish@lightening.com	$635.00
St. Elmo Steak House	Indianapolis	stelmo@aol.com	$420.00
Osetra The Fish House	San Diego	osetra@aol.com	$360.00
Bellagio	Las Vegas	bellagio@comcast.net	$210.00
The Park Hotel	Charlotte	parkhotel@zoom.net	$90.00

FIGURE 4-5 Query 3: Best Customers

Query 4: Add to Inventory

Create a query called Add to Inventory. This should be an update query that adds today's purchases from the suppliers to the current inventory. Do not run the query; simply set it up and save it.

Query 5: Subtract from Inventory

Create a query called Subtract from Inventory. This should be an update query that subtracts today's customer orders from the current inventory. Do not run the query; simply set it up and save it.

Report 1: Weekly Sales

Create a report called Weekly Sales. Your report's output should show column headings for Customer Name, Customer City, Product Name, Quantity, and Total Value. As part of this report, create a query that calculates the Total Value of each customer's order of each item for the current week. After you create the query, group the report on Customer Name. Adjust the output so that Customer City is on the same line as Customer Name. Clean up the report by deleting any extra total lines and making sure all data is in the correct format. Although your data will differ, your output should resemble the portion of the report shown in Figure 4-6.

Weekly Sales

Customer Name	Customer City	Product Name	Quantity	Total Value
Altamar	Miami			
		Xlarge	15	$825.00
Total Quantity and Value			15	*$825.00*
Bellagio	Las Vegas			
		Large	5	$210.00
Total Quantity and Value			5	*$210.00*
Big Fish	Dearborn			
		Xlarge	5	$275.00
		Large	5	$210.00
		Medium	5	$150.00
Total Quantity and Value			15	*$635.00*
Bound'Ry Restaurant	Nashville			
		Xlarge	15	$825.00
Total Quantity and Value			15	*$825.00*

FIGURE 4-6 Report 1: Weekly Sales

Report 2: Weekly Purchases

Create a report called Weekly Purchases. Your report's output should show headers for Supplier Name, Product Name, and Quantity. As part of this report, create a query that brings the data together for only the current week. After you create the query, group the report on Supplier Name. Clean up the report by deleting any extra total lines and making sure all data is in the correct format. Although your data will differ, your output should resemble that shown in Figure 4-7.

Weekly Purchases

Supplier Name	Product Name	Quantity
Falmouth Fish		
	Xlarge	25
	Xlarge	12
	Large	35
Total		72
Lunenburg Fisheries		
	Medium	27
	Large	10
	Xlarge	50
	Medium	20
Total		107
Grand Total		179

FIGURE 4-7 Report 2: Weekly Purchases

Navigation Pane: Easy Navigation

Create a custom group of objects in the navigation pane for easy access to the database. Include all of the objects in the database and categorize them as Data, Forms, Queries, and Reports. Call the custom group Easy Navigation. Its interface should resemble that shown in Figure 4-8. *Note:* The items under Data are purposely hidden here because they would reveal the database design.

FIGURE 4-8 Navigation Pane: Easy Navigation

ASSIGNMENT 3: MAKING A PRESENTATION

Create a presentation for JITL. Pay particular attention to the needs of those who might use the database and are not familiar with Microsoft Access. Your presentation should take less than 15 minutes, including a brief question-and-answer period.

DELIVERABLES

1. Word-processed design of tables
2. Tables created in Access
3. Form: Orders
4. Query 1: Customers in a Specified City
5. Query 2: Most Popular Product
6. Query 3: Best Customers
7. Query 4: Add to Inventory (not printed)
8. Query 5: Subtract from Inventory (not printed)
9. Report 1: Weekly Sales
10. Report 2: Weekly Purchases
11. Navigation Pane: Easy Navigation (not printed)
12. Any other required tutorial printouts or tutorial CD

Staple all pages together. Put your name and class number at the top of the page. Make sure your disk or CD is labeled if required.

THE CRAB SHACK EMPLOYEE DATABASE

Designing a Relational Database to Create Tables, Forms, Queries, a Report, and a Navigation Pane

PREVIEW

In this case, you'll design a relational database for employees of a beach-themed restaurant. After your database design is completed and correct, you will create database tables and populate them with data. Then you will produce two forms, four queries, one report, and one navigation pane.

The forms will serve two purposes: (1) allow employees to record their work hours and (2) allow management to assign dormitory rooms for the summer help.

The queries will:

1. Allow management to find out which employees fill a specific job in case an employee who is scheduled for that job calls in sick and cannot work.
2. Allow management to see which workers were late for their shift.
3. Allow management to prove to a licensing authority that certain workers are over 18 years of age.
4. Calculate the workers' pay, including overtime.

The report will list the job types and their respective pay and whether they require certification. The navigation pane will allow easy access to the data, forms, queries, and report.

PREPARATION

Before attempting this case, you should:

- Have some experience designing databases and using Microsoft Access.
- Complete any part of Database Design Tutorial A that your instructor assigns.
- Complete any part of Access Tutorial B that your instructor assigns or refer to the tutorial as necessary.
- Refer to Tutorial F as necessary.

BACKGROUND

The popularity of the beach as a vacation destination has helped the Crab Shack restaurant and bar become a popular hangout for both young and old. The Crab Shack is located on the bay side of a beach resort town in Maryland. During the winter months, the Crab Shack serves very few patrons; but in the summer, the place is crowded with vacationers. Despite its name, the Crab Shack is actually a large and trendy establishment that can serve hundreds of patrons.

To keep the customers amused, the owners trucked in tons of sand to make a portion of the property near the water resemble a tropical island. The owners planted a dozen large palm trees and added tables and benches so patrons could enjoy their food and drinks on the "beach." The restaurant also features an area where boats can tie up to buoys and a spot where customers can lounge in the water either at high tables or on floating tubes. In addition to serving great Maryland steamed crabs, the Crab Shack hires two bands and two DJs for the summer to keep the atmosphere lively.

This coming summer the Crab Shack is expanding, and the owners want to keep employee information on a computer. Until now, all employee information has been recorded and stored on paper. You are asked to design and implement an employee database before the busy summer season begins.

The Design of the Database

You must keep a number of parameters in mind when designing this employee database. The owners want to keep track of their seasonal employees by storing employee information such as name, address, telephone number, Social Security number, and date of birth. Because temporary employees are bunked in dormitories, employee gender is noted for dormitory assignments. In addition, each employee is assigned a unique employee number so that identity theft does not become an issue, as could be the case if Social Security numbers were used. That unique number is used to record working hours and to issue paychecks.

There are many jobs at the Crab Shack, and each employee has a particular assigned job. At the restaurant operation end, there are cooks, servers, bartenders, and security workers. At the water area, the Crab Shack assigns certified lifeguards to watch patrons who are floating at the high tables or on tubes. Certified boat handlers meet patrons arriving by boat and then bring the customers to the beach area. Musicians and DJs are employed directly by the Crab Shack for the summer season. In addition, the Crab Shack hires a large number of gardeners to groom the beach area by grading it with extra sand, planting and maintaining the area's numerous palm trees, and picking up trash.

The Business Owners' Needs

The Crab Shack's owners have conveyed to you their business needs, which are as follows:

Forms

The owners would like employees to record on a computer form the hours they worked.

In addition, this bustling restaurant has its own dormitory for housing out-of-state seasonal workers. Those workers are assigned dorm rooms based on gender and birth date. Each worker is placed in a double room with another worker of his or her age and gender. The owners would like to be able to assign dormitory rooms using a second form.

Queries

There are a number of challenges to running this business during the busy summer season. One challenge is filling shifts at the last minute. When an employee calls in sick, the manager usually needs to fill that position immediately. The manager would like to be able to run a query that prompts for a type of job and get output that shows all of the employees who are suited to perform a particular job.

Another challenge is keeping track of which employees arrive late for work. Employee tardiness is an ongoing problem for management. Thus, the manager would like to monitor which workers are late for their shifts by typing in the date, time, and job of a particular shift and seeing a list of employees who are late.

The local town authorities require that all food and beverage handlers at the Crab Shack be over 18 years of age; but lifeguards, boat handlers, and gardeners can be under 18. Often the authorities stop by and check employee records to make sure the Crab Shack is complying with this law. The owner would like a query that displays the workers who are over 18 years old from today's date.

An additional query that the payroll manager would like to run is one that calculates the monthly pay for each worker. This would include overtime pay, which is one-and-a-half times the regular rate for any employee who works more than 8 hours in a 24-hour time period.

Report

Finally, a simple report that displays the job types and their associated salaries and whether an employee needs certification (e.g., to serve as a lifeguard) needs to be created to enable management to run job ads in local newspapers and on the Web.

Navigation Pane

You also would like to create a navigation pane to allow management and workers to access the data, forms, queries, and reports easily.

ASSIGNMENT 1: CREATING THE DATABASE DESIGN

In this assignment, you design your database tables on paper, using a word processing program. Pay close attention to the tables' logic and structure. Do not start your Access code (Assignment 2) before getting feedback from your instructor on Assignment 1. Keep in mind that you will need to look at the requirements for Assignment 2 to design your fields and tables properly. Good programming practice is to look at the required outputs before designing your database. When designing the database, observe the following guidelines:

- Determine the tables you'll need by listing on paper the name of each table and the fields that it should contain. Avoid data redundancy. Do not create a field if it can be created by a "calculated field" in a query.
- Plan on at least one transaction table. Avoid duplicating data.
- Plan on a logical field.
- Document your tables by using the Table facility of your word processor. Your word-processed tables should resemble the format of the table in Figure 5-1.
- Mark the appropriate key field(s). You can designate a key field by inserting an asterisk (*) next to the field name. Keep in mind that some tables need a compound primary key to uniquely identify a record within a table.
- Print the database design.

Table Name	
Field Name	Data Type (text, numeric, currency, etc.)
...	...
...	...

FIGURE 5-1 Table design

> **NOTE**
>
> Have your design approved before beginning Assignment 2; otherwise, you may need to redo Assignment 2.

ASSIGNMENT 2: CREATING THE DATABASE AND MAKING FORMS, QUERIES, A REPORT, AND A NAVIGATION PANE

In this assignment, you create database tables in Access and populate them with data. Then you create two forms, four queries, a report, and a navigation pane.

Assignment 2A: Creating Tables in Access

In this part of the assignment, you create your tables in Access. Use the following guidelines:

- Using your classmates as the workers, type records into the tables. Make sure you enter addresses. Include at least two workers for each type of job.
- Assume there are nine jobs: lifeguard, security guard, bartender, server, cook, boat handler, musician, DJ, and gardener.
- Create a realistic pay-per-hour scale. Assume that employees are paid twice a month—on the 15th and the last day of the month.
- Have some workers work more than 8 hours in a 24-hour period. Enter all hours worked as being in the second half of May.
- Although you need to create a table for the dorm assignments, do not assign any dorms. (That can be done in the form.) Assume all rooms are double rooms and are divided between male and female.
- Appropriately limit the size of the text fields; for example, a ZIP Code does not require the default setting of 255 characters in length.
- Print all tables.

Assignment 2B: Creating Forms, Queries, a Report, and a Navigation Pane

There are two forms, four queries, one report, and one navigation pane to create, as outlined in the background of this case.

Form 1: Employee Hours Worked

Create a form with a subform that allows employees to record their hours worked. Include all employee information along with the information about hours worked. Save the form as Employee Hours Worked. Your data will vary; but when you view a record, it should resemble that shown in Figure 5-2.

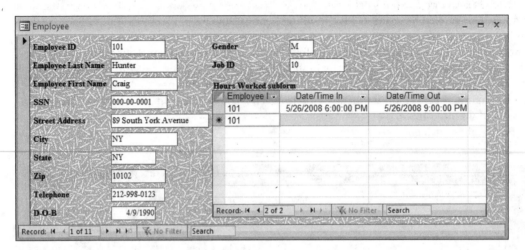

FIGURE 5-2   Form 1: Employee Hours Worked

Form 2: Employee Dorm Assignment

Create a second form called Employee Dorm Assignment with a subform that shows the employee information along with the dorm assignment information, as shown in Figure 5-3. (Your data will differ.) Do not create a relationship between the tables for this form and subform. You should be able to scroll through the records and see what rooms are available for each gender and then manually assign rooms.

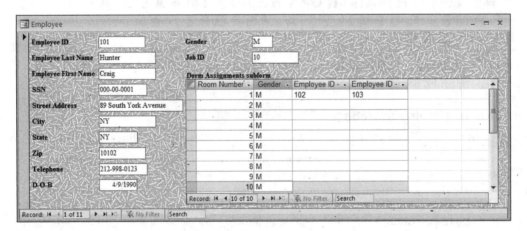

FIGURE 5-3   Form 2: Employee Dorm Assignment

Query 1: Employees by Specified Type

Create a query called Employees by Specified Type. This query should allow the user to prompt for a particular job title. Display the Employee ID, Employee Last Name, Employee First Name, Telephone, and Job Title. If you were to input the job Lifeguard into the query, your output would resemble that shown in Figure 5-4, although it would include two entries (as specified by the table-creating guidelines in Assignment 2A) and the data in these entries would vary.

Employees by Specified Type

Employee ID	Employee Last Name	Employee First Name	Telephone	Job Title
101	Hunter	Craig	212-998-0123	Lifeguard

FIGURE 5-4   Query 1: Employees by Specified Type

Query 2: Late Employees

Create a query called Late Employees. Display Employee Last Name, Employee First Name, Telephone, Date/Time In (for the time the employee started work), and the Job Title. Your query should ask for two inputs: the date and time the job was supposed to begin and the job title. For example, if the job was set to begin at 9 PM on 5/26/08, you would input that information into your query. Your query should output all of the records that begin after that specified time. Assume the manager will run this query during the current shift to show the names of workers who are currently late. Your data will differ, but your output should resemble that shown in Figure 5-5.

Late Employees

Employee Last Name	Employee First Name	Telephone	Date/Time In	Job Title
Fleisher	Emily	301-765-4523	5/26/2008 9:05:00 PM	Bartender

FIGURE 5-5   Query 2: Late Employees

Query 3: Employees over 18

Create a query called Employees over 18. Display Employee Last Name, Employee First Name, SSN (Social Security number) and D-O-B (date of birth). Your query should make sure that each of the employees it outputs is 18 years of age from today's date. Do not hard-code today's date. Use the Access function that calls the current date. Your data will differ, but the output should resemble that shown in Figure 5-6.

Employees over 18

Employee Last Name	Employee First Name	SSN	D-O-B
Richards	Bernie	000-00-0010	9/5/1981
Cressman	Anita	000-00-0011	1/3/1982
Fleisher	Emily	000-00-0002	12/19/1982
Fleisher	Peggy	000-00-0003	12/19/1982
John	Sanchez	000-00-0004	9/10/1970
West	Johnny	000-00-0005	4/9/1981
West	Mary	000-00-0006	12/15/1985
Cool	DJ	000-00-0007	4/1/1975
Trent	Barbara	000-00-0008	11/2/1979
Merit	Arthur	000-00-0009	8/30/1985

FIGURE 5-6   Query 3: Employees over 18

Query 4: May's Second Pay

Create a query called May's Second Pay. Display Employee Last Name, Employee First Name, SSN (Social Security number), and Pay. Calculate the pay for the second half of May only. You will need to use an If statement in the query design to figure out how to handle overtime pay. Data will differ, but your output should look like that shown in Figure 5-7. (*Note:* Only a portion of the output is shown.)

May's Second Pay

Employee Last Name	Employee First Name	SSN	Pay
Cool	DJ	000-00-0007	$187.00
Cressman	Anita	000-00-0011	$54.00
Fleisher	Emily	000-00-0002	$53.25
Fleisher	Peggy	000-00-0003	$33.00
Hunter	Craig	000-00-0001	$36.00
John	Sanchez	000-00-0004	$165.75
Merit	Arthur	000-00-0009	$94.50
Richards	Bernie	000-00-0010	$82.00
Trent	Barbara	000-00-0008	$154.00
West	Johnny	000-00-0005	$310.00
West	Mary	000-00-0006	$310.00

FIGURE 5-7 Query 4: May's Second Pay

Report: Job Types

Create a report called Job Types. Your report's output should show headers for Job ID, Job Title, and Salary/Hour and whether Certification is Required. Your output should resemble that shown in Figure 5-8.

Job Types

Job ID	*Job Title*	*Salary/Hour*	*Certification Required?*
10	Lifeguard	$12.00	☑
11	Security Guard	$10.50	☑
12	Bartender	$9.00	☑
13	Server	$8.25	☐
14	Cook	$10.25	☐
15	Boat Handler	$9.00	☑
16	Gardener	$12.75	☐
17	Musician	$20.00	☐
18	DJ	$22.00	☐

FIGURE 5-8 Report: Job Types

Navigation Pane

Create a navigation pane to access the data, forms, queries, and reports easily. Call your navigation pane "Easy Navigation."

ASSIGNMENT 3: MAKING A PRESENTATION

Create a presentation for the Crab Shack. Show employees how they can record the hours they worked. Train the manager to use the form to assign dorm rooms. Your presentation should take less than 15 minutes, including a brief question-and-answer period.

DELIVERABLES

1. Word-processed design of tables
2. Tables created in Access
3. Form 1: Employee Hours Worked
4. Form 2: Employee Dorm Assignment
5. Query 1: Employees by Specified Type
6. Query 2: Late Employees
7. Query 3: Employees over 18
8. Query 4: May's Second Pay
9. Report: Job Types
10. Navigation Pane: Easy Navigation
11. Presentation materials
12. Any other required tutorial printouts or tutorial CD

Staple all pages together. Put your name and class number at the top of the page. Make sure your disk or CD is labeled if required.

PART 2

BUILDING A DECISION SUPPORT SYSTEM IN EXCEL

A **decision support system (DSS)** is a computer program that can represent, either mathematically or symbolically, a problem that a user needs to solve. Such a representation is, in effect, a model of a problem.

Here's how a DSS program works: The DSS program accepts input from the user or looks at data in files on a disk. Then the DSS program runs the input and any other necessary data through the model. The program's output is the information the user needs to solve a problem. Some DSS programs recommend a solution to a problem.

A DSS can be written in any programming language that lets a programmer represent a problem. For example, a DSS can be built in a third-generation language such as Visual Basic or in a database package such as Access. A DSS also can be written in a spreadsheet package such as Excel.

The Excel spreadsheet package has standard built-in arithmetic functions as well as many statistical and financial functions. Thus, many kinds of problems—such as those in accounting, operations, and finance—can be modeled in Excel.

This tutorial has the following four sections:

1. **Spreadsheet and DSS Basics:** In this section, you'll learn how to create a DSS program in Excel. Your program will be in the form of a cash flow model. This section will give you practice in spreadsheet design and in building a DSS program.
2. **Scenario Manager:** In this section, you'll learn how to use an Excel tool called the Scenario Manager. With any DSS package, one problem with playing "what if" is this: Where do you physically record the results from running each set of data? Typically, a user writes the inputs and related results on a sheet of paper. Then—ridiculously enough—the user might have to input the data *back* into a spreadsheet for further analysis. The Scenario Manager solves that problem. It can be set up to capture inputs and results as "scenarios," which are then summarized on a separate sheet in the Excel workbook.
3. **Practice Using Scenario Manager:** You will work on a new problem, a case using the Scenario Manager.
4. **Review of Excel Basics:** This brief section reviews the information you'll need to do the spreadsheet cases that follow this tutorial.

SPREADSHEET AND DSS BASICS

Assume it is late 2007 and you are trying to build a model of what a company's net income (profit) and cash flow will be the next two years (2008 and 2009). This is the problem: to forecast net income and cash flow in those years. The company is likely to use the forecasts to answer a business question or make a strategic decision, so the estimates should be as accurate as you can make them. After researching the problem, you decide that the estimates should be based on three things: (1) 2007 results, (2) estimates of the underlying economy, and (3) the cost of products the company sells.

Your model will use an income statement and cash flow framework. The user can input values for two possible states of the economy in 2008–2009: an *O* for an optimistic outlook or a *P* for a pessimistic outlook. The state of the economy is expected to affect the number of units the company can sell as well as each unit's selling price. In a good, or *O*, economy, more units can be sold at a higher price. The user also can input values into your model for two possible cost-of-goods-sold price directions: a *U* for up or a *D* for down. A *U* means that the cost of an item sold will be higher than it was in 2007; a *D* means that the cost will be less.

Presumably, the company will do better in a good economy and with lower input costs—but how much better? Such relationships are too complex for most people to assess in their head, but a software model can assess them easily. Thus, the user can play "what if" with the input variables and note the effect on net income and year-end cash levels. For example, a user can ask these questions: What if the economy is good and costs go up? What will net income and cash flow be in that case? What will happen if the economy is down and costs go down? What will be the company's net income and cash flow in that case? With an Excel software model available, the answers to those questions are easily quantified.

Organization of the DSS Model

Your spreadsheets should have the following sections, which will be noted in boldfaced type throughout this tutorial and in the Excel cases featured in this textbook:

- **CONSTANTS**
- **INPUTS**
- **SUMMARY OF KEY RESULTS**
- **CALCULATIONS** (of values that will be used in the **INCOME STATEMENT AND CASH FLOW STATEMENT**)
- **INCOME STATEMENT AND CASH FLOW STATEMENT**

Here as an extended illustration, a DSS model is built for the forecasting problem described. Next, you'll look at each spreadsheet section. Figure C-1 and Figure C-2 show how to set up the spreadsheet.

	A	B	C	D
1	**TUTORIAL EXERCISE**			
2				
3	**CONSTANTS**	**2007**	**2008**	**2009**
4	TAX RATE	NA	0.33	0.35
5	NUMBER OF BUSINESS DAYS	NA	300	300
6				
7	**INPUTS**	**2007**	**2008**	**2009**
8	ECONOMIC OUTLOOK (O = OPTIMISTIC; P = PESSIMISTIC)	NA		NA
9	PURCHASE-PRICE OUTLOOK (U = UP; D = DOWN)	NA		NA
10				
11	**SUMMARY OF KEY RESULTS**	**2007**	**2008**	**2009**
12	NET INCOME AFTER TAXES	NA		
13	END-OF-THE-YEAR CASH ON HAND	NA		
14				
15	**CALCULATIONS**	**2007**	**2008**	**2009**
16	NUMBER OF UNITS SOLD IN A DAY	1000		
17	SELLING PRICE PER UNIT	7.00		
18	COST OF GOODS SOLD PER UNIT	3.00		
19	NUMBER OF UNITS SOLD IN A YEAR	NA		

FIGURE C-1   Tutorial skeleton 1

	A	B	C	D
21	**INCOME STATEMENT AND CASH FLOW STATEMENT**	**2007**	**2008**	**2009**
22	BEGINNING-OF-THE-YEAR CASH ON HAND	NA		
23				
24	SALES (REVENUE)	NA		
25	COST OF GOODS SOLD	NA		
26	INCOME BEFORE TAXES	NA		
27	INCOME TAX EXPENSE	NA		
28	NET INCOME AFTER TAXES	NA		
29				
30	END-OF-THE-YEAR CASH ON HAND (BEGINNING-OF-THE-YEAR CASH, PLUS NET INCOME AFTER TAXES)	10000		

FIGURE C-2   Tutorial skeleton 2

Each spreadsheet section is discussed next.

NOTE

Some of the numbers in the spreadsheet shown here have no decimals, whereas some have two decimals. Decimal precision can be controlled using the decimal icons in the Home tab's Number group. That formatting procedure is discussed more fully in the Review of Excel Basics section of this tutorial.

95

The CONSTANTS Section

This section of Figure C-1 records values that are used in spreadsheet calculations. In a sense, the constants are inputs, except that they do not change. In this tutorial, constants are TAX RATE and the NUMBER OF BUSINESS DAYS.

The INPUTS Section

The inputs shown in Figure C-1 are for the ECONOMIC OUTLOOK and PURCHASE-PRICE OUTLOOK (manufacturing input costs). Inputs could conceivably be entered for *each year* the model is covered (here, 2008 and 2009). That would let you enter an O for 2008's economy in one cell and a P for 2009's economy in another cell. Alternatively, one input for the two-year period could be entered in one cell. For simplicity, this tutorial uses the latter approach.

The SUMMARY OF KEY RESULTS Section

This section of the spreadsheet captures 2008 and 2009 NET INCOME AFTER TAXES (profit) for the year and END-OF-THE-YEAR CASH ON HAND, which you should assume are the two relevant outputs of this model. The summary merely repeats, in one easy-to-see place, results that appear in otherwise widely spaced places in the spreadsheet. (It also makes for easier graphing of results at a later time.)

The CALCULATIONS Section

This area is used to compute the following data:

1. The NUMBER OF UNITS SOLD IN A DAY, which is a function of the 2007 value and the economic outlook input
2. The SELLING PRICE PER UNIT, which is similarly derived
3. The COST OF GOODS SOLD PER UNIT, which is a function of the 2007 value and of purchase-price outlook
4. The NUMBER OF UNITS SOLD IN A YEAR, which equals the number of units sold in a day times the number of business days

Those formulas could be embedded in the **INCOME STATEMENT AND CASH FLOW STATEMENT** section of the spreadsheet, which will be described shortly. Doing that, however, would result in expressions that are complex and difficult to understand. Putting the intermediate calculations into a separate **CALCULATIONS** section breaks up the work into modules. That is good form because it simplifies your programming.

The INCOME STATEMENT AND CASH FLOW STATEMENT Section

This section is the "body" of the spreadsheet. It shows the:

1. BEGINNING-OF-THE-YEAR CASH ON HAND, which equals cash at the end of the *prior* year.
2. SALES (REVENUE), which equals the units sold in the year times the unit selling price.
3. COST OF GOODS SOLD, which is units sold in the year times the price paid to acquire or make the unit sold.
4. INCOME BEFORE TAXES, which equals sales, less total costs.
5. INCOME TAX EXPENSE, which is zero when there are losses; otherwise, it is the income before taxes times the tax rate. (INCOME TAX EXPENSE is sometimes called INCOME TAXES.)
6. NET INCOME AFTER TAXES, which equals income before taxes, less income tax expense.
7. END-OF-THE-YEAR CASH ON HAND, which is beginning-of-the-year cash on hand plus net income. (In the real world, cash flow estimates must account for changes in receivables and payables. In this case, assume sales are collected immediately—that is, there are no receivables or bad debts. Also assume that suppliers are paid immediately—that is, there are no payables.)

Construction of the Spreadsheet Model

Next, you will work through the following three steps to build your spreadsheet model:

1. Make a "skeleton" of the spreadsheet and call it **TUTC.xls**.
2. Fill in the "easy" cell formulas.
3. Enter the "hard" spreadsheet formulas.

Make a Skeleton

Your first step is to set up a skeleton worksheet. The worksheet should have headings, text string labels, and constants—but no formulas.

To set up the skeleton, you must first grasp the problem *conceptually*. The best way to do that is to work *backward* from what the "body" of the spreadsheet will look like. Here the body is the **INCOME STATEMENT AND CASH FLOW STATEMENT** section. Set that up in your mind or on paper; then do the following:

- Decide what amounts should be in the **CALCULATIONS** section. In the income statement of this tutorial's model, SALES (revenue) will be NUMBER OF UNITS SOLD IN A DAY times SELLING PRICE PER UNIT. You will calculate the intermediate amounts (NUMBER OF UNITS SOLD IN A YEAR and SELLING PRICE PER UNIT) in the **CALCULATIONS** section.
- Set up the **SUMMARY OF KEY RESULTS** section by deciding what *outputs* are needed to solve the problem. The **INPUTS** section should be reserved for amounts that can change—the controlling variables—which are the ECONOMIC OUTLOOK and the PURCHASE-PRICE OUTLOOK.
- Use the **CONSTANTS** section for values you will need to use but are not in doubt; that is, you will not have to input them or calculate them. Here the TAX RATE is a good example of such a value.

AT THE KEYBOARD

Type in the Excel skeleton shown in Figure C-1 and Figure C-2.

> **NOTE**
>
> A designation of *NA* means that a cell will not be used in any formula in the worksheet. The 2007 values are needed only for certain calculations; so for the most part, the 2007 column's cells show *NA*. (Recall that the forecast is for 2008 and 2009.) Also be aware that you can "break" a text string in a cell by pressing the Alt and Enter keys at the same time at the break point. That makes the cell "taller." Formatting of cells to show centered data and creation of borders is discussed at the end of this tutorial.

Fill in the "Easy" Formulas

The next step in building a spreadsheet model is to fill in the "easy" formulas. The cells affected (and what you should enter) are discussed next.

To prepare, you should format the cells in the **SUMMARY OF KEY RESULTS** section for no decimals. (Formatting for numerical precision is discussed at the end of this tutorial.) The **SUMMARY OF KEY RESULTS** section is shown in Figure C-3. As previously mentioned, the **SUMMARY OF KEY RESULTS** section simply echoes results shown in other places. Consider Figure C-1 and Figure C-2. Note that C28 in Figure C-2 holds the NET INCOME AFTER TAXES. The idea here is for you to echo that amount in C12 of Figure C-1. So the formula in C12 is =C28. Translation: "Copy what is in C28 into C12." It's that simple.

> **NOTE**
>
> With the insertion point in C12, the contents of that cell—in this case, the formula =C28—show in the editing window, which is above the lettered column indicators, as shown in Figure C-3.

C12		f_x =C28		
	A	B	C	D
11	**SUMMARY OF KEY RESULTS**	2007	2008	2009
12	NET INCOME AFTER TAXES	NA	0	
13	END-OF-THE-YEAR CASH ON HAND	NA		

FIGURE C-3   Echo 2008 NET INCOME AFTER TAXES

At this point, C28 is empty (and thus has a zero value), but that does not prevent you from copying. So copy cell C12's formula to the right of cell D12. (The copy operation does not actually "copy.") Copying puts =D28 into D12, which is what you want. (Year 2009's NET INCOME AFTER TAXES is in D28.)

To perform the Copy operation, use the following steps:

1. Select (click in) the cell (or range of cells) that you want to copy. That activates the cell (or range) for copying.
2. Press the Control key and *C* at the same time (Ctrl+C).
3. Select the destination cell. (If a range of cells is the destination, select the upper left cell of the destination range.)
4. Press the Control key and *V* at the same time (Ctrl+V).
5. Press the Escape key to deactivate the copied cell (or range).

Another way to copy is as follows:

1. Select the Home tab.
2. Select (click in) the cell (or range of cells) that you want to copy.
3. In the Clipboard group, select Copy.
4. Select the destination cell. (If a range of cells is the destination, select the upper left cell of the destination range.)
5. In the Clipboard group, select Paste.
6. Press the Escape key to deactivate the copied cell (or range).

As you can see in Figure C-4, END-OF-THE-YEAR CASH ON HAND for 2007 cash is in cell C13. Echo the cash results in cell C30 to cell C13. (Put =C30 in cell C13, as shown in Figure C-4.) Copy the formula from C13 to D13.

C13		f_x =C30		
	A	B	C	D
11	**SUMMARY OF KEY RESULTS**	2007	2008	2009
12	NET INCOME AFTER TAXES	NA	0	0
13	END-OF-THE-YEAR CASH ON HAND	NA	0	

FIGURE C-4   Echo 2008 END-OF-THE-YEAR CASH ON HAND

At this point, the **CALCULATIONS** section formulas will not be entered because they are not all "easy" formulas. Move on to the easier formulas in the **INCOME STATEMENT AND CASH FLOW STATEMENT** section, as if the calculations were already done. Again, the fact that the **CALCULATIONS** section cells are empty does not stop you from entering formulas in this section. You should format the cells in the **INCOME STATEMENT AND CASH FLOW STATEMENT** section for zero decimals.

As you can see in Figure C-5, BEGINNING-OF-THE-YEAR CASH ON HAND is the cash on hand at the end of the *prior* year. In C22 for 2008, type =B30. The "skeleton" you just entered is shown in Figure C-5. Cell B30 has the END-OF-THE-YEAR CASH ON HAND for 2007.

C22	▼	f_x	=B30

	A	B	C	D
21	INCOME STATEMENT AND **CASH FLOW STATEMENT**	**2007**	**2008**	**2009**
22	BEGINNING-OF-THE-YEAR CASH ON HAND	NA	10000	
23				
24	SALES (REVENUE)	NA		
25	COST OF GOODS SOLD	NA		
26	INCOME BEFORE TAXES	NA		
27	INCOME TAX EXPENSE	NA		
28	NET INCOME AFTER TAXES	NA		
29				
30	END-OF-THE-YEAR CASH ON HAND (BEGINNING-OF-THE-YEAR CASH, PLUS NET INCOME AFTER TAXES)	10000		

FIGURE C-5   Echo of END-OF-THE-YEAR CASH ON HAND for 2007 to BEGINNING-OF-THE-YEAR CASH ON HAND for 2008

Figure C-6 shows the next step, which is to copy the formula in cell C22 to the right. SALES (REVENUE) is NUMBER OF UNITS SOLD IN A YEAR times SELLING PRICE PER UNIT. In cell C24, enter =C17*C19, as shown in Figure C-6.

	A	B	C	D
15	**CALCULATIONS**	**2007**	**2008**	**2009**
16	NUMBER OF UNITS SOLD IN A DAY	1000		
17	SELLING PRICE PER UNIT	7.00		
18	COST OF GOODS SOLD PER UNIT	3.00		
19	NUMBER OF UNITS SOLD IN A YEAR	NA		
20				
21	INCOME STATEMENT AND **CASH FLOW STATEMENT**	**2007**	**2008**	**2009**
22	BEGINNING-OF-THE-YEAR CASH ON HAND	NA	10000	0
23				
24	SALES (REVENUE)	NA	0	
25	COST OF GOODS SOLD	NA		
26	INCOME BEFORE TAXES	NA		
27	INCOME TAX EXPENSE	NA		
28	NET INCOME AFTER TAXES	NA		
29				
30	END-OF-THE-YEAR CASH ON HAND (BEGINNING-OF-THE-YEAR CASH, PLUS NET INCOME AFTER TAXES)	10000		

FIGURE C-6   Enter the formula to compute 2008 SALES

The formula C17*C19 multiplies units sold for the year by the unit selling price. (Cells C17 and C19 are empty now, which is why SALES shows a zero after the formula is entered.) Copy the formula to the right to D24.

COST OF GOODS SOLD is handled similarly. In C25, type =C18*C19. That equals NUMBER OF UNITS SOLD IN A YEAR times COST OF GOODS SOLD PER UNIT. Copy the formula to the right.

In cell C26, the formula for INCOME BEFORE TAXES is =C24–C25. Enter the formula. Copy it to the right.

In the United States, income taxes are paid only on positive income before taxes. In cell C27 shown in Figure C-7, the INCOME TAX EXPENSE is zero when the INCOME BEFORE TAXES is zero or less; otherwise, INCOME TAX EXPENSE equals the TAX RATE times the INCOME BEFORE TAXES. The TAX RATE is a constant (in C4). An IF statement is needed to express this logic:

> IF(INCOME BEFORE TAXES is <= 0, put zero tax in C27,
> else in C27 put a number equal to multiplying the
> TAX RATE times the INCOME BEFORE TAXES)

C26 stands for the concept INCOME BEFORE TAXES, and C4 stands for the concept TAX RATE. So in Excel, substitute those cell addresses:

> =IF(C26 <= 0, 0, C4 * C26)

Copy the income tax expense formula to the right.

In cell C28, NET INCOME AFTER TAXES is INCOME BEFORE TAXES, less INCOME TAX EXPENSE: =C26–C27. Enter and copy the formula to the right.

The END-OF-THE-YEAR CASH ON HAND is BEGINNING-OF-THE-YEAR CASH ON HAND plus NET INCOME AFTER TAXES. In cell C30, enter =C22+C28. The **INCOME STATEMENT AND CASH FLOW STATEMENT** section at that point is shown in Figure C-7. Copy the formula to the right.

	A	B	C	D
21	**INCOME STATEMENT AND CASH FLOW STATEMENT**	**2007**	**2008**	**2009**
22	BEGINNING-OF-THE-YEAR CASH ON HAND	NA	10000	10000
23				
24	SALES (REVENUE)	NA	0	0
25	COST OF GOODS SOLD	NA	0	0
26	INCOME BEFORE TAXES	NA	0	0
27	INCOME TAX EXPENSE	NA	0	0
28	NET INCOME AFTER TAXES	NA	0	0
29				
30	END-OF-THE-YEAR CASH ON HAND (BEGINNING-OF-THE-YEAR CASH, PLUS NET INCOME AFTER TAXES)		10000	10000

FIGURE C-7 Status of INCOME STATEMENT AND CASH FLOW STATEMENT

Put in the "Hard" Formulas

The next step is to finish the spreadsheet by filling in the "hard" formulas.

AT THE KEYBOARD

In C8, enter an *O* for OPTIMISTIC; and in C9, enter *U* for UP. There is nothing magical about those particular values—they just give the worksheet formulas some input to process. Recall that the inputs will cover both 2008 and 2009. Make sure *NA* is in D8 and D9 just to remind yourself that those cells will not be used for input or by other worksheet formulas. Your **INPUTS** section should look like the one shown in Figure C-8.

	A	B	C	D
7	**INPUTS**	**2007**	**2008**	**2009**
8	ECONOMIC OUTLOOK (O = OPTIMISTIC; P = PESSIMISTIC)	NA	O	NA
9	PURCHASE-PRICE OUTLOOK (U = UP; D = DOWN)	NA	U	NA

FIGURE C-8 Entering two input values

Recall that cell addresses in the **CALCULATIONS** section are already referred to in formulas in the **INCOME STATEMENT AND CASH FLOW STATEMENT** section. The next step is to enter formulas for those calculations. Before doing that, format NUMBER OF UNITS SOLD IN A DAY and NUMBER OF UNITS SOLD IN A YEAR for zero decimals and format SELLING PRICE PER UNIT and COST OF GOODS SOLD PER UNIT for two decimals.

The easiest formula in the **CALCULATIONS** section is the NUMBER OF UNITS SOLD IN A YEAR, which is the calculated NUMBER OF UNITS SOLD IN A DAY (in C16) times the NUMBER OF BUSINESS DAYS (in C5). In C19, enter =C5*C16, as shown in Figure C-9.

C19		f_x	=C5*C16	
	A	B	C	D
1	**TUTORIAL EXERCISE**			
2				
3	CONSTANTS	2007	2008	2009
4	TAX RATE	NA	0.33	0.35
5	NUMBER OF BUSINESS DAYS	NA	300	300
6				
7	INPUTS	2007	2008	2009
8	ECONOMIC OUTLOOK (O = OPTIMISTIC; P = PESSIMISTIC)	NA	O	NA
9	PURCHASE-PRICE OUTLOOK (U = UP; D = DOWN)	NA	U	NA
10				
11	SUMMARY OF KEY RESULTS	2007	2008	2009
12	NET INCOME AFTER TAXES	NA	0	0
13	END-OF-THE-YEAR CASH ON HAND	NA	10000	10000
14				
15	CALCULATIONS	2007	2008	2009
16	NUMBER OF UNITS SOLD IN A DAY	1000		
17	SELLING PRICE PER UNIT	7.00		
18	COST OF GOODS SOLD PER UNIT	3.00		
19	NUMBER OF UNITS SOLD IN A YEAR	NA	0	

FIGURE C-9 Entering the formula to compute 2008 NUMBER OF UNITS SOLD IN A YEAR

Copy the formula to cell D19 for year 2009.

Assume that if the ECONOMIC OUTLOOK is OPTIMISTIC, the 2008 NUMBER OF UNITS SOLD IN A DAY will be 6 percent more than that in 2007; in 2009, they will be 6 percent more than that in 2008. Also assume that if the ECONOMIC OUTLOOK is PESSIMISTIC, the NUMBER OF UNITS SOLD IN A DAY in 2008 will be 1 percent less than those sold in 2007; in 2009, they will be 1 percent less than those sold in 2008. An IF statement is needed in C16 to express that idea:

> IF(economy variable = OPTIMISTIC,
> > then NUMBER OF UNITS SOLD IN A DAY will go UP 6%,
> > > else NUMBER OF UNITS SOLD IN A DAY will go DOWN 1%)

Substituting cell addresses:

> =IF($C$8 = "O", 1.06 * B16, 0.99 * B16)

N O T E

In Excel, quotation marks denote labels. The input is a one-letter label. So the quotation marks around the O are needed. Also note that multiplying by 1.06 results in a 6 percent increase, whereas multiplying by .99 results in a 1 percent decrease.

Enter the entire IF formula into cell C16, as shown in Figure C-10. Absolute addressing is needed ($C$8), because the address is in a formula that gets copied *and* you do not want the cell reference to change (to D8, which has the value *NA*) when you copy the formula to the right. Absolute addressing maintains the $C$8 reference when the formula is copied. Copy the formula in C16 to D16 for 2009.

C16		f_x	=IF($C$8="O",1.06*B16,0.99*B16)	
	A	B	C	D
15	CALCULATIONS	2007	2008	2009
16	NUMBER OF UNITS SOLD IN A DAY	1000	1060	
17	SELLING PRICE PER UNIT	7.00		
18	COST OF GOODS SOLD PER UNIT	3.00		

FIGURE C-10 Entering the formula to compute 2008 NUMBER OF UNITS SOLD IN A DAY

The SELLING PRICE PER UNIT is also a function of the ECONOMIC OUTLOOK. Assume the two-part rule is as follows:

- If the ECONOMIC OUTLOOK is OPTIMISTIC, the SELLING PRICE PER UNIT in 2008 will be 1.07 times that of 2007; in 2009, it will be 1.07 times that of 2008.
- On the other hand, if the ECONOMIC OUTLOOK is PESSIMISTIC, the SELLING PRICE PER UNIT in 2008 and 2009 will equal the per-unit price in 2007; that is, the price will not change.

Test your understanding of the selling price calculation by figuring out the formula for cell C17. Enter the formula and copy it to the right. You will need to use absolute addressing. (Can you see why?)

The COST OF GOODS SOLD PER UNIT is a function of the PURCHASE-PRICE OUTLOOK. Assume the two-part rule is as follows:

- If the PURCHASE-PRICE OUTLOOK is UP (U), COST OF GOODS SOLD PER UNIT in 2008 will be 1.25 times that of year 2007; in 2009, it will be 1.25 times that of 2008.
- On the other hand, if the PURCHASE-PRICE OUTLOOK is DOWN (D), the multiplier each year will be 1.01.

Again, to test your understanding, figure out the formula for cell C18. Enter and copy the formula to the right. You will need to use absolute addressing.

Your selling price and cost of goods sold formulas, given OPTIMISTIC and UP input values, should yield the calculated values shown in Figure C-11.

	A	B	C	D
		2007	2008	2009
15	CALCULATIONS			
16	NUMBER OF UNITS SOLD IN A DAY	1000	1060	1124
17	SELLING PRICE PER UNIT	7.00	7.49	8.01
18	COST OF GOODS SOLD PER UNIT	3.00	3.75	4.69
19	NUMBER OF UNITS SOLD IN A YEAR	NA	318000	337080

FIGURE C-11   Calculated values given OPTIMISTIC and UP input values

Assume you change the input values to PESSIMISTIC and DOWN. Your formulas should yield the calculated values shown in Figure C-12.

	A	B	C	D
		2007	2008	2009
15	CALCULATIONS			
16	NUMBER OF UNITS SOLD IN A DAY	1000	990	980
17	SELLING PRICE PER UNIT	7.00	7.00	7.00
18	COST OF GOODS SOLD PER UNIT	3.00	3.03	3.06
19	NUMBER OF UNITS SOLD IN A YEAR	NA	297000	294030

FIGURE C-12   Calculated values given PESSIMISTIC and DOWN input values

That completes the body of your spreadsheet. The values in the **CALCULATIONS** section ripple through the **INCOME STATEMENT AND CASH FLOW STATEMENT** section because the income statement formulas reference the calculations. Assuming inputs of OPTIMISTIC and UP, the income and cash flow numbers should look like those in Figure C-13.

	A	B	C	D
21	INCOME STATEMENT AND CASH FLOW STATEMENT	2006	2007	2008
22	BEGINNING-OF-THE-YEAR CASH ON HAND	NA	10000	806844
23				
24	SALES (REVENUE)	NA	2381820	2701460
25	COST OF GOODS SOLD	NA	1192500	1580063
26	INCOME BEFORE TAXES	NA	1189320	1121398
27	INCOME TAX EXPENSE	NA	392476	392489
28	NET INCOME AFTER TAXES	NA	796844	728909
29				
30	END-OF-THE-YEAR CASH ON HAND (BEGINNING-OF-THE-YEAR CASH, PLUS NET INCOME AFTER TAXES)	10000	806844	1535753

FIGURE C-13   Completed **Income Statement and Cash Flow Statement** section

SCENARIO MANAGER

You are now ready to use the Scenario Manager to capture inputs and results as you play "what if" with the spreadsheet.

Note that there are four possible combinations of input values: O-U (Optimistic-Up), O-D (Optimistic-Down), P-U (Pessimistic-Up), and P-D (Pessimistic-Down). Financial results for each combination will be different. Each combination of input values can be referred to as a "scenario." Excel's Scenario Manager records the results of each combination of input values as a separate scenario and then shows a summary of all scenarios in a separate worksheet. Those summary worksheet values can be used as a raw table of numbers and then printed or copied into a Word document. The table of data can then be the basis for an Excel chart, which also can be printed or inserted into a memo.

You have four possible scenarios for the economy and the purchase price of goods sold: Optimistic-Up, Optimistic-Down, Pessimistic-Up, and Pessimistic-Down. The four input-value sets produce different financial results. When you use the Scenario Manager, define the four scenarios; then have Excel (1) sequentially run the input values "behind the scenes" and (2) put the results for each input scenario in a summary sheet.

When you define a scenario, you give the scenario a name and identify the input cells and input values. You do that for each scenario. Then you identify the output cells so Excel can capture the outputs in a summary sheet.

▣ AT THE KEYBOARD

To start, select the Data tab. The Data Tools group has a What-If Analysis icon. Select that drop-down arrow and click the Scenario Manager menu choice. Initially, no scenarios are defined; and Excel tells you that, as you can see in Figure C-14.

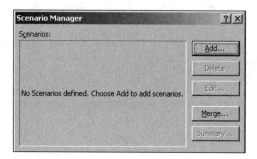

FIGURE C-14   Initial Scenario Manager window

With this window, you are able to add, delete, or change (edit) scenarios. Toward the end of the process, you also can create the summary sheet.

NOTE

When working with this window and its successors, do not press the Enter key to navigate. Use mouse clicks to move from one step to the next.

To continue with defining a scenario, click the Add button. In the resulting Add Scenario window, give the first scenario a name: OPT-UP. Then type the input cells in the Changing cells window—here, C8:C9. (*Note:* C8 and C9 are contiguous input cells. Noncontiguous input cell ranges are separated by a comma.) Excel may add dollar signs to the cell address, but do not be concerned about that. The window should look like the one shown in Figure C-15.

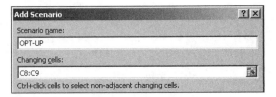

FIGURE C-15   Entering OPT-UP as a scenario

Now click **OK**, which moves you to the Scenario Values window. Here you indicate what the INPUT *values* will be for the scenario. The values in the cells *currently* in the spreadsheet will be displayed. They might or might not be correct for the scenario you are defining. For the OPT-UP scenario, you need to enter an *O* and a *U*, if not the current values. Enter those values if needed, as shown in Figure C-16.

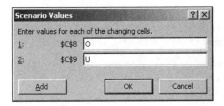

FIGURE C-16   Entering OPT-UP scenario input values

Click **OK**, which takes you back to the Scenario Manager window. You can now enter the other three scenarios, following the same steps. Do that now. Enter the OPT-DOWN, PESS-UP, and PESS-DOWN scenarios and related input values. After doing that, you should see that the names and changing cells for the four scenarios have been entered, as in Figure C-17.

FIGURE C-17   Scenario Manager window with all scenarios entered

You are now able to create a summary sheet that shows the results of running the four scenarios. Click the Summary button. You'll get the Scenario Summary window. You must provide Excel with the output cell addresses—they will be the same for all four scenarios. (The output *values* in those output cells change as input values are changed, but the addresses of the output cells do not change.)

Assume you are interested primarily in the results that have accrued at the end of the two-year period. Those are your two 2009 **SUMMARY OF KEY RESULTS** section cells for NET INCOME AFTER TAXES and END-OF-THE-YEAR CASH ON HAND (D12 and D13). Type those addresses in the window's input area, as shown in Figure C-18. (*Note:* If result cells are noncontiguous, the address ranges can be entered, separated by a comma.)

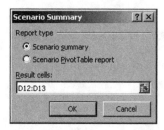

FIGURE C-18 Entering Result cells addresses in Scenario Summary window

Then click **OK**. Excel runs each set of inputs and collects results as it goes. (You do not see that happening on the screen.) Excel makes a *new* sheet, titled the "Scenario Summary" (denoted by the sheet's lower tab) and takes you there, as shown in Figure C-19.

	A	B	C	D	E	F	G	H
1								
2		**Scenario Summary**						
3			Current Values:		OPT-UP	OPT-DOWN	PESS-UP	PESS-DOWN
5		**Changing Cells:**						
6		$C$8	O	O	O	P	P	
7		$C$9	U	U	D	U	D	
8		**Result Cells:**						
9		$D$12		728909	728909	1085431	441964	752953
10		$D$13		1535753	1535753	2045679	1098681	1552944
11		Notes: Current Values column represents values of changing cells at						
12		time Scenario Summary Report was created. Changing cells for each						
13		scenario are highlighted in gray.						

FIGURE C-19 Scenario Summary sheet created by Scenario Manager

One somewhat annoying visual element is that the Current Values in the spreadsheet are given an output column. That duplicates one of the four defined scenarios. You can delete that extra column by following these steps: (1) Select the Home tab; (2) within the Cells group, select the Delete icon's drop-down arrow; and (3) select Delete Sheet Columns.

NOTE

A sheet's row can be deleted the same way as a column can be deleted. You select Delete Sheet Rows at the end.

Another annoyance is that column A goes unused. You can click and delete it as you've been doing to move everything to the left. That should make columns of data easier to see on the screen without scrolling. Other ways to make the worksheet easier to read include: (1) entering words in column A to describe the input and output cells; (2) centering cell values by using the Center Text icon in the Home tab's Alignment group; and (3) showing data in Currency format by using the Number Format drop-down menu within the Home tab's Number group.

When you're done, your summary sheet should resemble the one shown in Figure C-20.

	A	B	C	D	E	F
1	Scenario Summary					
2			OPT-UP	OPT-DOWN	PESS-UP	PESS-DOWN
4	Changing Cells:					
5	ECONOMIC OUTLOOK	$C$8	O	O	P	P
6	PURCHASE PRICE OUTLOOK	$C$9	U	D	U	D
7	Result Cells (2009):					
8	NET INCOME AFTER TAXES	$D$12	$728,909	$1,085,431	$441,964	$752,953
9	END-OF-THE-YEAR CASH ON HAND	$D$13	$1,535,753	$2,045,679	$1,098,681	$1,552,944

FIGURE C-20 Scenario Summary sheet after formatting

Note that column C shows the OPTIMISTIC-UP case. NET INCOME AFTER TAXES in that scenario is $728,909, and END-OF-THE-YEAR CASH ON HAND is $1,535,753. Columns D, E, and F show the other scenario results.

Here is an important postscript to this exercise: DSS spreadsheets are used to guide decision making. That means that the spreadsheet's results must be interpreted in some way. Here are a few practice questions based on the results in Figure C-20: With that data, what combination of year 2009 NET INCOME AFTER TAXES and END-OF-THE-YEAR CASH ON HAND would be best? Clearly, OPTIMISTIC-DOWN (O-D) is the best result, right? It yields the highest income and highest cash. What is the worst combination? PESSIMISTIC-UP (P-U), right? It yields the lowest income and lowest cash.

Results are not always that easy to interpret, but the analytical method is the same. You have a complex situation that you cannot understand very well without software assistance. You build a model of the situation in the spreadsheet, enter the inputs, collect the results, and then interpret the results to help with decision making.

Summary Sheets

When you do Scenario Manager spreadsheet case studies, you'll need to manipulate summary sheets and their data. Now you will look at some of those operations.

Rerunning the Scenario Manager

The Scenario Summary sheet does not update itself when the spreadsheet formulas or inputs change. To see an updated Scenario Summary sheet, the user must rerun the Scenario Manager. To rerun the Scenario Manager, click the Summary button in the Scenario Manager dialog box; then click the OK button. That makes another summary sheet. It does not overwrite a prior one.

Deleting Unwanted Scenario Manager Summary Sheets

Suppose you want to delete a summary sheet. With the summary sheet on the screen, follow this procedure: (1) Select the Home tab; (2) within the Cells group, select the Delete icon's drop-down arrow; and (3) select Delete Sheet. Note that you will be asked if you really mean to delete this sheet. Press Delete if you do.

Charting Summary Sheet Data

The summary sheet results can be conveniently charted using the Chart Wizard. Charting Excel data is discussed in Tutorial E.

Copying Summary Sheet Data to the Clipboard

You may want to put the summary sheet data into the Clipboard to use in a Word document. To do that, follow these steps:

1. Select the data range.
2. Copy the data range by following the copying operation described earlier in this tutorial. That puts the data range into the clipboard.
3. Open your Word document.
4. Click the cursor where you want the upper left part of the graphic to be positioned.
5. Paste the data into the document by selecting Paste in the Home tab's Clipboard group.

PRACTICE USING SCENARIO MANAGER

Suppose you have an uncle who works for a large company. He has a good job and makes a decent salary (currently, $80,000 a year). At age 65, he can retire from his company, which will be in 2014. He will start drawing his pension then.

However, the company has an early-out plan. Under the plan, the company asks employees to quit (called "preretirement"). The company then pays those employees a bonus in the year they quit and each year thereafter, up to the official retirement date, which is through the year 2013 for your uncle. Then employees start to receive their actual pension—in your uncle's case, in 2014. This early-out program would let your uncle leave the company before 2014. Until then, he could find a part-time hourly job to make ends meet and then leave the workforce entirely in 2014.

The opportunity to leave early is open through 2013. That means your uncle could stay with the company in 2008, then leave the company any year in the period 2009 to 2013 and get the early-out bonuses in the years he is retired. If he retires in 2009, he would lose the 2008 bonus. If he retires in 2010, he would lose the 2008 and 2009 bonuses. That would continue all the way through 2013.

Another factor in your uncle's thinking is whether to continue his country club membership. He likes the club, but it is a cash drain. The early-out decision can be assessed each year, but the decision about the country club membership must be made now; if your uncle does not withdraw in 2008, he says he will remain a member (and incur costs) through 2013.

Your uncle has called you in to make a Scenario Manager spreadsheet model of his situation. Your spreadsheet would let him play "what if" with the preretirement and country club possibilities and see his various projected personal finance results for 2008–2013. With each scenario, your uncle wants to know what "cash on hand" will be available for each year in the period.

Complete the spreadsheet for your uncle. Your **SUMMARY OF KEY RESULTS, CALCULATIONS**, and **INCOME STATEMENT AND CASH FLOW STATEMENT** section cells must show values *by cell formula*. That is, in those areas, do not hard-code amounts. Also do not use the address of a cell if its contents are *NA* in any of your formulas. Set up your spreadsheet skeleton as shown in the figures that follow. Name your spreadsheet **UNCLE.xls**.

CONSTANTS Section

Your spreadsheet should have the constants shown in Figure C-21. An explanation of line items follows the figure.

	A	B	C	D	E	F	G	H
1	**YOUR UNCLE'S EARLY RETIREMENT DECISION**							
2	CONSTANTS	2007	2008	2009	2010	2011	2012	2013
3	SALARY INCREASE FACTOR	NA	0.03	0.03	0.02	0.02	0.01	0.01
4	PART TIME WAGES EXPECTED	NA	10000	10200	10500	10800	11400	12000
5	BUY OUT AMOUNT	NA	30000	25000	20000	15000	5000	0
6	COST OF LIVING (NOT RETIRED)	NA	41000	42000	43000	44000	45000	46000
7	COUNTRY CLUB DUES	NA	12000	13000	14000	15000	16000	17000

FIGURE C-21 CONSTANTS section values

- SALARY INCREASE FACTOR: Your uncle's salary at the end of 2007 will be $80,000. As you can see, raises are expected each year; for example, a 3 percent raise is expected in 2008. If your uncle does not retire, he will get his salary and small raise for the year.
- PART TIME WAGES EXPECTED: Your uncle has estimated his part-time wages as if he were retired and working part-time in the 2008–2013 period.
- BUY OUT AMOUNT: The amounts for the company's preretirement buyout plan are shown. If your uncle retires in 2008, he gets $30,000, $25,000, $20,000, $15,000, $5,000, and zero in the years 2008 to 2013, respectively. If he leaves in 2009, he will give up the $30,000 2008 payment, but will get $25,000, $20,000, $15,000, $5,000, and zero in the years 2009 to 2013, respectively.
- COST OF LIVING (NOT RETIRED): Your uncle has estimated how much cash he needs to meet his living expenses, assuming he continues to work for the company. His cost of living would be $41,000 in 2008, increasing each year thereafter.
- COUNTRY CLUB DUES: Country club dues are $12,000 for 2008. They increase each year thereafter.

INPUTS Section

Your spreadsheet should have the inputs shown in Figure C-22. An explanation of line items follows the figure.

	A	B	C	D	E	F	G	H
9	INPUTS	2007	2008	2009	2010	2011	2012	2013
10	RETIRED [R] or WORKING [W]	NA						
11	STAY IN CLUB? [Y] OR [N]	NA		NA	NA	NA	NA	NA

FIGURE C-22 INPUTS section

- RETIRED OR WORKING: Enter an *R* if your uncle retires in a year or a *W* if he is still working. If he is working through 2013, the pattern **WWWWWW** should be entered. If his retirement is in 2008, the pattern **RRRRRR** should be entered. If he works for three years and then retires in 2011, the pattern **WWWRRR** should be entered.
- STAY IN CLUB?: If your uncle stays in the club from 2008–2013, a **Y** should be entered. If your uncle leaves the club in 2008, an **N** should be entered. The decision applies to all years.

SUMMARY OF KEY RESULTS Section

Your spreadsheet should show the results in Figure C-23.

	A	B	C	D	E	F	G	H
13	SUMMARY OF KEY RESULTS	2007	2008	2009	2010	2011	2012	2013
14	END-OF-THE-YEAR CASH ON HAND	NA						

FIGURE C-23 SUMMARY OF KEY RESULTS section

Each year's END-OF-THE-YEAR CASH ON HAND value is echoed from cells in the spreadsheet body.

CALCULATIONS Section

Your spreadsheet should calculate, by formula, the values shown in Figure C-24. Calculated amounts are used later in the spreadsheet. An explanation of line items follows the figure.

	A	B	C	D	E	F	G	H
16	CALCULATIONS	2007	2008	2009	2010	2011	2012	2013
17	TAX RATE	NA						
18	COST OF LIVING	NA						
19	YEARLY SALARY OR WAGES	80000						
20	COUNTRY CLUB DUES PAID	NA						

FIGURE C-24 CALCULATIONS section

- TAX RATE: Your uncle's tax rate depends on whether he is retired. Retired people have lower overall tax rates. If he is retired in a year, your uncle's rate is expected to be 15 percent of income before taxes. In a year in which he works full-time, the rate will be 30 percent.
- COST OF LIVING: In any year that your uncle continues to work for the company, his cost of living is the amount shown in COST OF LIVING (NOT RETIRED) in the **CONSTANTS** section in Figure C-21. But if he chooses to retire, his cost of living is $15,000 less than the amount shown in the figure.
- YEARLY SALARY OR WAGES: If your uncle keeps working, his salary increases each year. The year-to-year percentage increases are shown in the **CONSTANTS** section. Thus, salary earned in 2008 would be more than that earned in 2007, salary earned in 2009 would be more than that earned in 2008, etc. If your uncle retires in a certain year, he will make the part-time wages shown in the **CONSTANTS** section.
- COUNTRY CLUB DUES PAID: If your uncle leaves the club, the dues are zero each year; otherwise, the dues are as shown in the **CONSTANTS** section.

The INCOME STATEMENT AND CASH FLOW STATEMENT Section

This section begins with the cash on hand at the beginning of the year. That is followed by the income statement, concluding with the calculation of cash on hand at the end of the year. The format is shown in Figure C-25. An explanation of line items follows the figure.

	A	B	C	D	E	F	G	H
22	INCOME STATEMENT AND CASH FLOW STATEMENT	2007	2008	2009	2010	2011	2012	2013
23	BEGINNING-OF-THE-YEAR CASH ON HAND	NA						
24								
25	SALARY OR WAGES	NA						
26	BUY OUT INCOME	NA						
27	TOTAL CASH INFLOW	NA						
28	COUNTRY CLUB DUES PAID	NA						
29	COST OF LIVING	NA						
30	TOTAL COSTS	NA						
31	INCOME BEFORE TAXES	NA						
32	INCOME TAX EXPENSE	NA						
33	NET INCOME AFTER TAXES	NA						
34								
35	END-OF-THE-YEAR CASH ON HAND (BEGINNING-OF-THE-YEAR CASH, PLUS NET INCOME AFTER TAXES)	30000						

FIGURE C-25   INCOME STATEMENT AND CASH FLOW STATEMENT section

- BEGINNING-OF-THE-YEAR CASH ON HAND: This is the END-OF-THE-YEAR CASH ON HAND at the end of the prior year.
- SALARY OR WAGES: This is a yearly calculation, which can be echoed here.
- BUY OUT INCOME: This is the year's buyout amount if your uncle is retired that year.
- TOTAL CASH INFLOW: This is the sum of salary or part-time wages and buyout amounts.
- COUNTRY CLUB DUES PAID: This is a calculated amount.
- COST OF LIVING: This is a calculated amount.
- TOTAL COSTS: These outflows are the sum of the COST OF LIVING and COUNTRY CLUB DUES PAID.
- INCOME BEFORE TAXES: This amount is the TOTAL CASH INFLOW, less TOTAL COSTS (outflows).
- INCOME TAX EXPENSE: This amount is zero when INCOME BEFORE TAXES is zero or less; otherwise, the calculated tax rate is applied to the INCOME BEFORE TAXES.
- NET INCOME AFTER TAXES: This is INCOME BEFORE TAXES, less TAX EXPENSE.
- END-OF-THE-YEAR CASH ON HAND: This is the BEGINNING-OF-THE-YEAR CASH plus the year's NET INCOME AFTER TAXES.

Scenario Manager Analysis

Set up the Scenario Manager and create a Scenario Summary sheet. Your uncle wants to look at the following four possibilities:

- Retire in 2008, staying in the club ("Loaf-In")
- Retire in 2008, leaving the club ("Loaf-Out")
- Work three more years and retire in 2011, staying in the club ("Delay-In")
- Work three more years and retire in 2011, leaving the club ("Delay-Out")

You can enter noncontiguous cell ranges as follows: C20..F20, C21, C22 (cell addresses are examples). The output cell should be only the 2013 END-OF-THE-YEAR CASH ON HAND cell.

Your uncle will choose the option that yields the highest 2013 END-OF-THE-YEAR CASH ON HAND. You must look at your Scenario Summary sheet to see which strategy yields the highest amount.

To check your work, you should attain the values shown in Figure C-26. (You can use the labels that Excel gives you in the left-most column or change the labels, as was done in Figure C-26.)

	A	B	C	D	E	F
1	Scenario Summary					
2			LOAF-IN	LOAF-OUT	DELAY-IN	DELAY-OUT
4	Changing Cells:					
5	RETIRE OR WORK, 2008	$C$10	R	R	W	W
6	RETIRE OR WORK, 2009	$D$10	R	R	W	W
7	RETIRE OR WORK, 2010	$E$10	R	R	W	W
8	RETIRE OR WORK, 2011	$F$10	R	R	R	R
9	RETIRE OR WORK, 2012	$G$10	R	R	R	R
10	RETIRE OR WORK, 2013	$H$10	R	R	R	R
11	IN CLUB, 2008-2013?	$C$11	Y	N	Y	N
12	Result Cells:					
13	END-OF-THE-YEAR CASH ON HAND	$H$14	-$68,400	$15,195	$8,389	$83,689

FIGURE C-26 Scenario Summary

REVIEW OF EXCEL BASICS

In this section, you'll begin by reviewing how to perform some basic operations. Then you'll work through more cash flow calculations. Reading and working through this section will help you complete the spreadsheet cases in this book.

Basic Operations

To begin, you'll review the following topics: formatting cells, showing Excel cell formulas, understanding circular references, using the And and the Or functions in IF statements, and using nested IF statements.

Formatting Cells

You may have noticed that some data in this tutorial's first spreadsheet was centered in the cells. Here is how to perform that operation:

1. Highlight the cell range to format.
2. Select the Home tab.
3. In the Alignment group, select the Middle Align icon to change the vertical alignment.
4. In the Alignment group, select the Center icon to change the horizontal alignment.

It is also possible to put a border around cells. That treatment might be desirable for highlighting **INPUTS** section cells. To perform that operation:

1. Highlight the cell that needs a border.
2. Select the Home tab.
3. In the Font group, select the drop-down arrow of the Bottom Border icon.
4. Choose the desired border's menu choice; Outside Border and Thick Box Border would be best for your purposes.

You can format numerical values for Currency by following these steps:

1. Highlight the cell or range of cells that should be formatted.
2. Select the Home tab.
3. In the Number group, use the Number Format drop-down arrow to select Currency.

You can format numerical values for decimal places using this procedure:

1. Highlight the cell or range of cells that should be formatted.
2. Select the Home tab.
3. In the Number group, click the Increase Decimal icon once to add one decimal value. Click the Decrease Decimal icon to eliminate a decimal value.

Showing Excel Cell Formulas

If you want to see Excel cell formulas, follow these steps:

1. Press the **Ctrl** key and the back quote key (') at the same time. (The back quote orients from northwest to southeast—on most keyboards, it is to the left of the exclamation point and shares the key with the diacritical tilde mark.)
2. To restore, press the **Ctrl** and (') back quote keys again.

Understanding a Circular Reference

A formula has a circular reference when the reference *refers to the cell the formula is in*. Excel cannot properly evaluate such a formula. The problem is best described by an example. Suppose the formula in cell C18 is =C18–C17. Excel is trying to compute a value for cell C18. To do that, it must evaluate the formula, then put the result on the screen in C18. Excel tries to subtract what is in C17 from what is in C18, but nothing is in C18. Can you see the circularity? To establish a value for C18, Excel must know what is in C18. The process is circular; hence, the term *circular reference*. In the example, the formula in C18 refers to C18. Excel points out circular references, if any exist, when you choose Open for a spreadsheet. Excel also points out circular references as you insert them during the building of a spreadsheet. The software alerts the user to circular references by opening at least one Help window and by drawing arrows between cells involved in the offending formula. You can close the windows, but that will not fix the situation. You must fix the formula that has the circular reference if you want the spreadsheet to give you accurate results.

Using the AND Function and the OR Function in IF Statements

An IF statement has the following syntax:

IF(test condition, result if test is True, result if test is False)

The test conditions in this tutorial's IF statements tested only one cell's value. But a test condition can test more than one value of a cell.

Here is an example from this tutorial's first spreadsheet in which selling price was a function of the economy. Assume, for the sake of illustration, 2008's selling price per unit depends on the economy *and* on the purchase-price outlook. There are two possibilities: (1) If the economic outlook is optimistic *and* the company's purchase-price outlook is down, the selling price will be 1.10 times the prior year's price. (2) In all other cases, the selling price will be 1.03 times the prior year's price. The first possibility's test requires two things to be true *at the same time*: C8 = "O" *AND* C9 = "D." To implement the test, the AND() function is needed. The code in cell C17 would be as follows:

=IF(AND($C$8 = "O", $C$9 = "D"), 1.10 * B17, 1.03 * B17)

When the test that uses the AND() function evaluates to True, the result is 1.10 * B17. When the test that uses the AND() function evaluates to False, the result is the second possibility's outcome: 1.03 * B17.

Now suppose the first possibility is as follows: If the economic outlook is optimistic *or* the purchase-price outlook is down, the selling price will be 1.10 times the prior year's price. Assume in all other cases the selling price will be 1.03 times the prior year's price. Now the test requires *only one of* two things to be true: C8 = "O" *OR* C9 = "D." To implement that, the OR() function would be needed. The code in cell C17 would be:

=IF(OR($C$8 = "O", $C$9 = "D"), 1.10 * B17, 1.03 * B17)

Using IF Statements Inside IF Statements

Recall from the previous section that an IF statement has this syntax:

IF(test condition, result if test is True, result if test is False)

In the examples shown thus far, only two courses of action were possible; so only one test was needed in the IF statement. There can, however, be more than two courses of action; when that is the case, the "result if test is False" clause needs to show further testing. Look at the following example.

Assume again that the 2008 selling price per unit depends on the economy and the purchase-price outlook. Here is the logic: (1) If the economic outlook is optimistic *and* the purchase-price outlook is down, the selling price will be 1.10 times the prior year's price. (2) If the economic outlook is optimistic *and* the purchase-price outlook is up, the selling price will be 1.07 times the prior year's price. (3) In all other cases, the selling price will be 1.03 times the prior year's price. The code in cell C17 would be as follows:

=IF(AND($C$8 = "O", $C$9 = "D"), 1.10 * B17,
    IF(AND($C$8 = "O", $C$9 = "U"), 1.07 * B17, 1.03 * B17))

The first IF statement tests to see if the economic outlook is optimistic and the purchase-price outlook is down. If not, further testing is needed to see whether the economic outlook is optimistic and the purchase-price outlook is up or whether some other situation prevails.

NOTE

Note the following:

- The line is broken in the previous example because the page is not wide enough; but in Excel, the formula would appear on one line.

- The embedded ìIFî is not preceded by an equal sign.

Example: Borrowing and Repayment of Debt

The Scenario Manager cases in this book require you to account for money that the company borrows or repays. Borrowing and repayment calculations are discussed next. At times, you will be asked to think about a question and fill in the answer. Correct responses are found at the end of this section.

To work through the Scenario Manager cases that follow, you must assume two things about a company's borrowing and repayment of debt. First, assume the company wants to have a certain minimum cash level at the end of a year (and thus at the start of the next year). Second, assume a bank will provide a loan to reach the minimum cash level if year-end cash falls short of the desired level.

Here are some numerical examples to test your understanding. Assume NCP stands for "net cash position" and equals beginning-of-the-year cash plus net income after taxes for the year. In other words, the NCP is the cash available at year end, before any borrowing or repayment. For the three examples in Figure C-27, compute the amounts the company needs to borrow to reach its minimum year-end cash level.

Example	NCP	Minimum Cash Required	Amount to Borrow
1	50,000	10,000	?
2	8,000	10,000	?
3	−20,000	10,000	?

FIGURE C-27 Examples of borrowing

One additional assumption you can make is that the company will use its excess cash at year end to pay off as much debt as possible without going below the minimum-cash threshold. Excess cash is the NCP *less* the minimum cash required on hand—amounts over the minimum are available to repay any debt.

In the examples shown in Figure C-28, compute excess cash and then compute the amount to repay. To aid your understanding, you also may want to compute ending cash after repayments.

Example	NCP	Minimum Cash Required	Beginning-of-the-Year Debt	Repay	Ending Cash
1	12,000	10,000	4,000	?	?
2	12,000	10,000	10,000	?	?
3	20,000	10,000	10,000	?	?
4	20,000	10,000	0	?	?
5	60,000	10,000	40,000	?	?
6	−20,000	10,000	10,000	?	?

FIGURE C-28 Examples of repayment

In this section's Scenario Manager cases, your spreadsheet will need two bank financing sections beneath the **INCOME STATEMENT AND CASH FLOW STATEMENT** section: The first section will calculate any needed borrowing or repayment at year's end to compute year-end cash. The second section will calculate the amount of debt owed at the end of the year, after borrowing or repayment of debt.

The first new section, in effect, extends the end-of-year cash calculation, which was shown in Figure C-13. Previously, the amount equaled cash at the beginning of the year plus the year's net income. Now the calculation will include cash obtained by borrowing and cash repaid. Figure C-29 shows the structure of the calculation.

	A	B	C	D
30	NET CASH POSITION (NCP) BEFORE BORROWING AND REPAYMENT OF DEBT (BEGINNING-OF-THE-YEAR CASH PLUS NET INCOME AFTER TAXES	NA		
31	PLUS: BORROWING FROM BANK	NA		
32	LESS: REPAYMENT TO BANK	NA		
33	EQUALS: END-OF-THE-YEAR CASH ON HAND	10000		

FIGURE C-29   Calculation of END-OF-THE-YEAR CASH ON HAND

The heading in cell A30 was previously END-OF-THE-YEAR CASH ON HAND. But BORROWING increases cash and REPAYMENT OF DEBT decreases cash. So END-OF-THE-YEAR CASH ON HAND is now computed two rows down (in C33 for year 2008 in the example). The value in row 30 must be a subtotal for the BEGINNING-OF-THE-YEAR CASH ON HAND plus the year's NET INCOME AFTER TAXES. That subtotal is called the NET CASH POSITION (NCP) BEFORE BORROWING AND REPAYMENT OF DEBT. (*Note:* Previously, the formula in cell C22 for BEGINNING-OF-THE-YEAR CASH ON HAND was =B30. Now that formula is =B33. It is copied to the right, as before, for the next year.)

The second new section computes end-of-year debt and is called DEBT OWED. That second new section is shown in Figure C-30.

	A	B	C	D
		2007	**2008**	**2009**
35	DEBT OWED			
36	BEGINNING-OF-THE-YEAR DEBT OWED	NA		
37	PLUS: BORROWING FROM BANK	NA		
38	LESS: REPAYMENT TO BANK	NA		
39	EQUALS: END-OF-THE-YEAR DEBT OWED	15000		

FIGURE C-30   DEBT OWED section

As you can see in Figure C-30, at the end of 2007, $15,000 was owed. END-OF-THE-YEAR DEBT OWED equals the BEGINNING-OF-THE-YEAR DEBT OWED plus any new BORROWING FROM BANK (which increases debt owed), less any REPAYMENT TO BANK (which reduces it). So in the example, the formula in cell C39 would be:

=C36+C37–C38

Assume the amounts for BORROWING FROM BANK and REPAYMENT TO BANK are calculated in the first new section. Thus, the formula in cell C37 would be =C31. The formula in cell C38 would be =C32. (BEGINNING-OF-THE-YEAR DEBT OWED is equal to the debt owed at the end of the prior year, of course. The formula in cell C36 for BEGINNING-OF-THE-YEAR DEBT OWED would be an echoed formula. Can you see what it would be? It's an exercise for you to complete. *Hint:* The debt owed at the beginning of a year equals the debt owed at the end of the prior year.)

Now that you have seen how the borrowing and repayment data is shown, the logic of the borrowing and repayment formulas can be discussed.

Calculation of BORROWING FROM BANK

The logic of this in English is:

> If (cash on hand before financing transactions is greater than the
> minimum cash required, then borrowing is not needed;
> otherwise, borrow enough to get to the minimum).

Or (a little more precisely):

> If (NCP is greater than the minimum cash required,
> then BORROWING FROM BANK = 0; otherwise,
> borrow enough to get to the minimum).

Suppose the desired minimum cash at year end is $10,000 and that value is a constant in your spreadsheet's cell C6. Assume the NCP is shown in your spreadsheet's cell C30. The formula (getting closer to Excel) would be as follows:

IF(NCP > Minimum Cash, 0; otherwise, borrow enough to get to the minimum).

You have cell addresses that stand for NCP (cell C30) and Minimum Cash (C6). To develop the formula for cell C31, substitute the cell addresses for NCP and Minimum Cash. The harder logic is that for the "otherwise" clause. At this point, you should look ahead to the borrowing answers at the end of this section, Figure C-31. In Example 2, $2,000 was borrowed. Which cell was subtracted from which other cell to calculate that amount? Substitute cell addresses in the Excel formula for 2008's borrowing formula in cell C31:

=IF(>=   , 0 ,   –)

The answer is at the end of this section in Figure C-33.

Calculation of REPAYMENT TO BANK

The logic of this in English is:

IF(beginning of year debt = 0, repay 0 because nothing is owed, but
   IF(NCP is less than the min, repay zero, because you must *borrow*, but
      IF(extra cash equals or exceeds the debt, repay the whole debt,
         ELSE (to stay above the min, repay only the extra cash))))

Look at the following formula. Assume the repayment will be in cell C32. Assume debt owed at the beginning of the year is in cell C36 and minimum cash is in cell C6. Substitute cell addresses for concepts to complete the formula for 2008 repayment. (Clauses are on different lines because of page width limitations.)

=IF(= 0, 0,
   IF(<=   , 0,
      IF((–   ) >=,
         (–   )))).

The answer is at the end of this section in Figure C-34.

Answers to Questions about Borrowing and Repayment Calculations

Figure C-31 and Figure C-32 give the answers to the questions posed about borrowing and repayment calculations.

Example	NCP	Minimum Cash Required	Borrow	Comments
1	50,000	10,000	Zero	NCP > Min.
2	8,000	10,000	2,000	Need 2,000 to get to Min. 10,000 – 8,000
3	–20,000	10,000	30,000	Need 30,000 to get to Min. 10,000 – (–20,000)

FIGURE C-31   Answers to examples of borrowing

Example	NCP	Minimum Cash Required	Beginning-of- the-Year Debt	Repay	Ending Cash
1	12,000	10,000	4,000	2,000	10,000
2	12,000	10,000	10,000	2,000	10,000
3	20,000	10,000	10,000	10,000	10,000
4	20,000	10,000	0	0	20,000
5	60,000	10,000	40,000	40,000	20,000
6	–20,000	10,000	10,000	NA	NA

FIGURE C-32   Answers to examples of repayment

Some notes about the repayment calculations shown in Figure C-32 follow.

- In Examples 1 and 2, only $2,000 is available for debt repayment (12,000 – 10,000) to avoid going below the minimum cash.
- In Example 3, cash available for repayment is $10,000 (20,000 – 10,000); so all beginning debt can be repaid, leaving the minimum cash.
- In Example 4, no debt is owed; so no debt need be repaid.
- In Example 5, cash available for repayment is $50,000 (60,000 – 10,000); so all beginning debt can be repaid, leaving more than the minimum cash.
- In Example 6, no cash is available for repayment. The company must borrow.

Figure C-33 and Figure C-34 show the calculations for borrowing and repayment of debt.

```
=IF( C30 >= C6, 0, C6 – C30)
```

FIGURE C-33   Calculation of borrowing

```
=IF( C36 = 0, 0, IF( C30 <= C6, 0, IF( (C30 – C6) >= C36, C36, ( C30 – C6))))
```

FIGURE C-34   Calculation of repayment

Saving Files after Using Microsoft Excel

As you work, you should make a habit of saving your files periodically. The top of the screen should have a Quick Access Toolbar, and the default Excel installation puts the Save icon there. (The icon looks like a diskette.) You can save your work by clicking that icon. Or you can save by clicking the Office Button, which is the relatively large button in the upper left corner of the screen. Choose the Save menu choice. The first time you save, you will use the Save As window to specify the following:

- A drive (using the My Documents drop-down menu)
- A filename (using the File name text box)
- A file type (by default, Excel 2007 makes an .xlsx file; but you can save in the Excel 2003 format or in other formats)

You will see the Save As window every time you choose Save As (not Save) when saving.

At the end of your Excel session, save your work using this three-step procedure:

1. Save the file one last time.
2. Using the Office Button, choose Close. If saving to a secondary disk, make sure the disk is still in its drive when you close. Closing the file takes the work off the screen. Do not take your disk out of its drive with the work still on the screen before you close. If you violate that rule, you may lose your work.
3. Using the Office Button, choose Exit Excel. That should take you back to Windows.

TRANSIT STRIKE NEGOTIATIONS

Decision Support Using Excel

PREVIEW

Your city's transit authority is a for-profit corporation that runs the buses and commuter trains. The company is near bankruptcy. The contract with the unionized employees has run out and must be renegotiated in a way that will let the company recover. In this case, you will use the Excel Scenario Manager to model the financial implications of some key variables in the labor negotiations.

PREPARATION

Before attempting this case, you should:

- Review spreadsheet concepts discussed in class and/or in your textbook.
- Complete any exercises that your instructor assigns.
- Complete any part of Access Tutorial C that your instructor assigns or refer to the tutorial as necessary.
- Review file-saving procedures for Windows programs, which are discussed in Tutorial C.
- Refer to Tutorial F as necessary.

BACKGROUND

Your city has a contract with a corporation called the Transit Authority that runs the buses and trains into, out of, and within the city. The men and women who drive the corporation's buses and trains are unionized. Their labor contract is expiring, and negotiations for a new four-year contract are getting under way.

The Transit Authority is unique in that it is a private corporation that receives an annual subsidy from the federal government. In addition, the corporation's actions have political implications: when people cannot get around town, they become angry with the mayor and city council.

The Transit Authority is losing money and has a sizable debt. It may be possible to raise ticket prices a small amount, but that alone will not solve the company's financial problems. Labor costs are too high, and company management knows it cannot continue the business without getting help from the drivers' union in the upcoming labor negotiations.

The company's chief financial officer (CFO) tells you that the Transit Authority's financial fate depends on four key financial variables: ticket prices, salaries, benefits, and employee attrition. Each of those variables is discussed next.

1. Ticket Prices—Management thinks that the average ticket price can increase no more than a dime per year over the next four years; the riders and the politicians would not stand for more than that. On the other side of the issue, management thinks that a price decrease would stimulate ridership, which might result in an increase in revenue.
2. Salary—The drivers will want a salary increase each year, but management will want to hold the line and might try to get a reduction in salaries.
3. Benefits—Medical insurance and other benefits are very expensive. The drivers will want better benefits, which would increase the company's costs. Management will want the drivers to shoulder more of the cost of benefits. For example, drivers could pay for a higher percentage of the cost of prescription drugs, which would reduce the company's costs.

4. Employee Attrition—There are currently 5,000 drivers. About 50 drivers a year retire. Management can choose to replace the drivers, thus keeping the number of drivers the same; or management might decide not to replace them, thus allowing the number of drivers to decline each year. Reducing the number of drivers will entail cutting back on the number of routes and the times that services are offered.

The company's CFO has asked you to make a spreadsheet model of the situation. Certain negotiating scenarios are under consideration. Your spreadsheet will model them in the Scenario Manager.

ASSIGNMENT 1: CREATING A SPREADSHEET FOR DECISION SUPPORT

In this assignment, you produce a spreadsheet that models the business decision. In Assignment 2, you write a memo to the company's CFO about your analysis and recommendation. Finally, in Assignment 3, you prepare an oral presentation of your analysis and recommendation.

First, you create the spreadsheet model of the company's financial situation. The model is for the four years of the new contract, 2008–2011. You will be given instructions on how each spreadsheet section should be set up before you enter cell formulas. Your spreadsheet should include these sections:

- CONSTANTS
- INPUTS
- SUMMARY OF KEY RESULTS
- CALCULATIONS
- INCOME STATEMENT AND CASH FLOW STATEMENT
- DEBT OWED

A discussion of each section follows. The spreadsheet skeleton is available to you, so you need not type it in. To access the spreadsheet skeleton, go to your Data files. Select Case 6; then select TRANSIT.xlsx.

CONSTANTS Section

Your spreadsheet should include the constants shown in Figure 6-1. An explanation of the line items follows the figure.

	A	B	C	D	E	F
1	**TRANSIT LABOR NEGOTIATION**					
2						
3	CONSTANTS	2007	2008	2009	2010	2011
4	TAX RATE	NA	0.15	0.15	0.15	0.15
5	CASH NEEDED TO START NEXT YEAR	NA	$2,000,000	$2,000,000	$2,000,000	$2,000,000
6	FIXED COSTS	NA	$46,000,000	$47,000,000	$48,000,000	$49,000,000
7	INTEREST RATE ON DEBT	NA	0.06	0.06	0.06	0.06
8	FEDERAL SUBSIDY EXPECTED	NA	$35,000,000	$35,000,000	$35,000,000	$35,000,000

FIGURE 6-1 CONSTANTS section

- TAX RATE: The tax rate on income before taxes in 2008–2011 is expected to be 15 percent each year, as shown.
- CASH NEEDED TO START NEXT YEAR: The company's bankers say that the company must start each year with a certain amount of cash, as shown.
- FIXED COSTS: Fixed costs include the cost of administration and maintenance on buses and railroad equipment, as well as other costs. Such costs are expected to increase gradually each year, as shown.
- INTEREST RATE ON DEBT: The interest rate on debt owed to bankers is expected to be 6 percent each year.
- FEDERAL SUBSIDY EXPECTED: The federal government supports transportation in the nation's cities. The company expects to receive $35 million each year, which is a form of revenue to the company.

INPUTS Section

Your spreadsheet should include the inputs shown in Figure 6-2. An explanation of the line items follows the figure.

	A	B	C	D	E	F
10	INPUTS	2007	2008	2009	2010	2011
11	TICKET PRICE CHANGE (.05 = UP A NICKEL, -.05 = DOWN A NICKEL, 0 = NO CHANGE)		NA	NA	NA	NA
12	SALARY CHANGE (.XX)	NA				
13	BENEFITS CHANGE (.XX)	NA				
14	ALLOW ATTRITION? (YES/NO)		NA	NA	NA	NA

FIGURE 6-2   INPUTS section

- TICKET PRICE CHANGE: The spreadsheet user enters a value for the average change in ticket price applicable each year. Coin receptacles cannot handle pennies, so multiples of five (a nickel) are entered. For example, if a nickel increase was expected each year, the user would enter .05. The value applies to each of the four years. Thus, the entry .05 means that 2008's ticket prices are five cents higher than 2007's, 2009's are five cents higher than 2008's, etc. A decrease in the period is possible; and in that case, a minus sign is entered. For example, a nickel decrease would be indicated by the entry -.05. Management does not think that the average ticket prices should be increased more than a dime in any one year.
- SALARY CHANGE: The spreadsheet user enters a value for the average salary change each year. If salaries are expected to increase 1 percent each year, the sequence .01, .01, .01, and .01 would be entered for 2008–2011. Suppose management wants to give a 1 percent raise the first year, a 2 percent raise the second year, etc. In that case, the sequence .01, .02, .03, and .04 would be entered. Thus, 2008's average salary would be 1 percent higher than 2007's, 2009's average salary would be 2 percent higher than 2008's, etc. A decrease in a period is possible; and in that case, a minus sign is entered.
- BENEFITS CHANGE: The spreadsheet user enters a value for the average change in benefits cost each year. The computation follows the same logic as that for the salary change entry.
- ALLOW ATTRITION?: Fifty drivers a year retire. If the company will not hire substitutes, the user enters the word YES—i.e., the company will allow attrition to reduce the number of drivers. If the company will hire drivers to replace retirees, the user enters NO.

Your instructor may tell you to apply Conditional Formatting to the ticket price input cell so that out-of-bounds values are highlighted in some way. (For example, an out-of-bounds entry could show up in red type or in boldfaced type). If you will be using Conditional Formatting, your instructor may provide a tutorial on that topic or may ask you to refer to Excel Help.

SUMMARY OF KEY RESULTS Section

Your spreadsheet should display the results shown in Figure 6-3. An explanation of the line items follows the figure.

	A	B	C	D	E	F
16	SUMMARY OF KEY RESULTS	2007	2008	2009	2010	2011
17	NET INCOME AFTER TAXES	NA				
18	END-OF-THE-YEAR CASH ON HAND	NA				
19	END-OF-THE-YEAR DEBT OWED	NA				

FIGURE 6-3   SUMMARY OF KEY RESULTS section

For each year, your spreadsheet should show (1) net income after taxes, (2) cash on hand at the end of the year, and (3) debt owed at the end of the year to bankers, as shown in Figure 6-3. All of those values are computed elsewhere in the spreadsheet and should be echoed here. You should format all of those cells for zero decimals.

CALCULATIONS Section

You should calculate various intermediate results that will be used in the income statement and cash flow statement that follows. Calculations, as shown in Figure 6-4, are based on inputs and on year-end 2007 values. When called for, use absolute addressing properly. An explanation of the line items follows the figure.

	A	B	C	D	E	F
21	CALCULATIONS	2007	2008	2009	2010	2011
22	NUMBER OF DRIVERS	5000				
23	NUMBER OF RIDERS	NA				
24	AVERAGE TICKET PRICE	$3.00				
25	AVERAGE SALARY	$40,000				
26	AVERAGE BENEFITS	$20,000				

FIGURE 6-4   CALCULATIONS section

- NUMBER OF DRIVERS: If attrition is not allowed to take place, the number of drivers a year will equal the number of drivers the prior year. However, if attrition is allowed, the number of drivers a year will be 50 fewer than the number of drivers the preceding year. Whether attrition will be allowed is an input value.
- NUMBER OF RIDERS: This is a function of two factors: the number of drivers and the change in the average ticket price.

  1. Riders live along service routes. Drivers are needed to service a route. Thus, the number of routes that can be serviced is a function of the number of drivers. If drivers are cut, fewer routes will be serviced and fewer riders will be picked up. Experience has shown that a driver will, on average, pick up 20,000 riders per year.
  2. If the average price of a ride goes up, some people will stop riding the buses and trains; they will drive their own cars or find another alternative. But if the average price goes down, more people will ride buses and trains. The adjustment factor is expected to be (1 / (1 + ticket price change)). Example: Assume a five-cent increase in ticket prices and 5,000 drivers. The number of riders = 20,000 * 5,000 * (1 / (1 + .05)) = 95,238,095 riders during the year. Can you see that a five-cent decrease in ticket prices (and 5,000 drivers) would mean 105,263,158 riders during the year? The change in ticket price is an input. The number of drivers is a calculation.

- AVERAGE TICKET PRICE: The average ticket price is a function of the prior year's ticket price and the ticket price change, which is an input. Example: Ticket prices will increase a nickel per year. 2008's average ticket price will be five cents greater than 2007's, 2009's average ticket price will be five cents greater than 2008's, etc., for the remaining years.
- AVERAGE SALARY: The average driver salary is a function of the prior year's average salary and the year's salary change factor, which is an input. Example: Salaries will increase 1 percent in 2008. 2008's average salary will be 1 percent greater than 2007's, 2009's average salary will be an increase or a decrease from 2008's by 2009's salary change factor, etc, for the remaining years. Note that the salary change factors are inputs, and they can be negative values.
- AVERAGE BENEFITS: The average driver benefits cost is a function of the prior year's average benefits cost and the year's benefits change factor, which is an input. Example: Benefits will increase 1 percent in 2008. 2008's average benefits will be 1 percent greater than 2007's, 2009's average benefits will be an increase or a decrease from 2008's by 2009's benefits change factor, etc, for the remaining years. Note that benefits change factors are inputs, and they can be negative values.

INCOME STATEMENT AND CASH FLOW STATEMENT Section

The forecast for net income and cash flow starts with the cash on hand at the beginning of the year. That is followed by the income statement and concludes with the calculation of cash on hand at the end of the year. For readability, format cells in this section for zero decimals. Your spreadsheet should look like the ones shown in Figure 6-5 and Figure 6-6. A discussion of the line items follows each figure.

	A	B	C	D	E	F
28	INCOME STATEMENT AND CASH FLOW STATEMENT	2007	2008	2009	2010	2011
29	BEGINNING-OF-THE-YEAR CASH ON HAND	NA				
30						
31	TICKET REVENUE	NA				
32	FEDERAL SUBSIDY	NA				
33	TOTAL REVENUE	NA				
34	SALARY COST	NA				
35	BENEFITS COST	NA				
36	FIXED COSTS	NA				
37	TOTAL COSTS	NA				
38	INCOME BEFORE INTEREST AND TAXES	NA				
39	INTEREST EXPENSE	NA				
40	INCOME BEFORE TAXES	NA				
41	INCOME TAX EXPENSE	NA				
42	NET INCOME AFTER TAXES	NA				

FIGURE 6-5 INCOME STATEMENT AND CASH FLOW STATEMENT section

- BEGINNING-OF-THE-YEAR CASH ON HAND: This is the cash on hand at the end of the prior year.
- TICKET REVENUE: This is a function of the number of riders (one ticket per rider) and the average ticket price. Both values are calculations.
- FEDERAL SUBSIDY: This is a constant from Figure 6-1, which can be echoed here.
- TOTAL REVENUE: This is the sum of ticket revenue and the federal subsidy.
- SALARY COST: This is a function of the number of drivers and the average driver salary. Both values are calculations.
- BENEFITS COST: This is a function of the number of drivers and the average driver benefits cost. Both values are calculations.
- FIXED COSTS: This is a constant, which can be echoed here.
- TOTAL COSTS: This is the sum of salary, benefits, and fixed costs.
- INCOME BEFORE INTEREST AND TAXES: This is the difference between total revenue and total costs.
- INTEREST EXPENSE: This is a function of the debt owed at the start of the year and the annual interest rate on debt, a constant.
- INCOME BEFORE TAXES: This is income before interest and taxes, minus interest expense.
- INCOME TAX EXPENSE: This value is zero if the income before taxes is zero or negative. Otherwise, income tax expense is a function of the year's tax rate and the income before income taxes.
- NET INCOME AFTER TAXES: This is the difference between income before taxes and income tax expense.

Continuing the statement, line items for the year-end cash calculation are discussed. In Figure 6-6, Column B is for 2007, Column C is for 2008, etc. Year 2007 values are NA except for END-OF-THE-YEAR CASH ON HAND, which is $2,000,000.

	A	B	C	D	E	F
44	NET CASH POSITION (NCP) BEFORE BORROWING AND REPAYMENT OF DEBT (BEG OF YR CASH + NET INCOME)	NA				
45	ADD: BORROWING FROM BANK	NA				
46	LESS: REPAYMENT TO BANK	NA				
47	EQUALS: END-OF-THE-YEAR CASH ON HAND	$2,000,000				

FIGURE 6-6 END-OF-THE-YEAR CASH ON HAND section

- NET CASH POSITION (NCP): The NCP at the end of a year equals cash at the beginning of the year plus the year's net income after taxes.
- ADD: BORROWING FROM BANK: Assume the company's banker will lend the company enough money at year's end to get to the minimum cash needed to start the next year. If the NCP is less than this minimum, the company must borrow enough money to get to the minimum amount. Borrowing increases cash on hand, of course.

- LESS: REPAYMENT TO BANK: If the NCP is more than the minimum cash at the end of a year and debt is owed, the company must pay off as much debt as possible (but not so much that the cash falls below the minimum cash required to start the next year). Repayments reduce cash on hand, of course.
- EQUALS: END-OF-THE-YEAR CASH ON HAND: This equals the NCP plus any borrowings, less any repayments.

DEBT OWED Section

The body of your spreadsheet ends with a calculation of debt owed at year's end, as shown in Figure 6-7. An explanation of the line items follows the figure. Year 2007 values are NA, except that the company owes $45,000,000 at the end of that year.

	A	2007	2008	2009	2010	2011
49	DEBT OWED					
50	BEGINNING-OF-THE-YEAR DEBT OWED	NA				
51	ADD: BORROWING FROM BANK	NA				
52	LESS: REPAYMENT TO BANK	NA				
53	EQUALS: END-OF-THE-YEAR DEBT OWED	$45,000,000				

FIGURE 6-7 DEBT OWED section

- BEGINNING-OF-THE-YEAR DEBT OWED: Cash owed at the beginning of a year equals cash owed at the end of the prior year.
- ADD: BORROWING FROM BANK: This amount has been calculated elsewhere and can be echoed in this section. Borrowings increase debt owed.
- LESS: REPAYMENT TO BANK: This amount has been calculated elsewhere and can be echoed in this section. Repayments reduce debt owed.
- EQUALS: END-OF-THE-YEAR DEBT OWED: This equals the amount owed at the beginning of a year, plus borrowings during the year, less repayments during the year.

ASSIGNMENT 2: USING THE SPREADSHEET FOR DECISION SUPPORT

You now complete the case by (1) using the spreadsheet to gather the data needed to assess the impact of ticket price changes, salary changes, benefits changes, and attrition decisions and (2) documenting your findings and recommendation in a memo.

The Transit Authority has not been doing well financially. The CFO needs a way to improve company finances. The goals are to become profitable and to reduce the company's debt load. Reducing debt load means attaining a debt level that is less than the company's current debt of $45 million.

The key variables are ticket price changes, salary changes, benefits changes, and attrition. To enter labor negotiations, the CFO needs to know what values of those variables will lead to profits and debt reduction. The CFO also needs to know what values to avoid so the company won't incur more losses and debt.

The CFO envisions five scenarios corresponding to five possible negotiating positions the company may find itself in, and he needs to know the financial results for each scenario.

Assignment 2A: Using the Spreadsheet to Gather Data

You have built the spreadsheet to model the business situation. The CFO wants to know what the values for profitability, cash on hand, and debt owed would be in 2011 for each of the five scenarios. The scenarios are shown in the following notations:

- No Change: In this scenario, no changes would occur: no changes in ticket price, salary, and benefits and no attrition to reduce the number of drivers.
- Traditional: The traditional approach would be to keep workers happy by raising salaries and benefits and not allowing attrition to reduce the number of drivers. In this scenario, the Transit Authority would pass on added costs to the customer by raising prices as follows: raise average

ticket price five cents; raise salaries 1 percent in 2008, 2 percent in 2009, 3 percent in 2010, and 4 percent in 2011; raise benefits 2 percent each of the four years; and do not allow attrition to reduce the number of drivers.

- Hopeful: It would be nice if the company could persuade the drivers' union not to request salary and benefits increases and allow attrition to reduce the number of drivers. That would reduce costs. In addition, the Transit Authority would raise average ticket prices five cents.
- Draconian: The company might opt to play hardball with the union—and with the riders. In this scenario, the Transit Authority would raise average ticket prices ten cents and give drivers a 1 percent raise in salary each year of the upcoming four years—but it also would reduce benefits by 2 percent each of those years. In addition, the Transit Authority would allow attrition to reduce the number of drivers.
- Lower Price: Would reducing ticket prices really help? In this scenario, the Transit Authority would drop average ticket prices by five cents, offer no salary increase each year, reduce benefits by 2 percent each year, and allow attrition to reduce the number of drivers.

The financial results in each scenario will act as a negotiating guide to management. You will use Scenario Manager to run "what-if" scenarios with the five sets of input values. See Tutorial C for the Scenario Manager procedures. Set up the five scenarios. The changing cells are the cells used to input changes in ticket price, salary, and benefits and the attrition decision. (*Note:* In the Scenario Manager, you can enter noncontiguous cell ranges as follows: C20..F20, C21, C22. Cell addresses cited here are arbitrary.) The output cells are the 2011 Summary of Key Results cells. Run the Scenario Manager to gather the data in a report. When you are done, print the spreadsheet with the inputs for any one of the scenarios and print the Scenario Manager summary sheet. Then save the spreadsheet file one last time (Office Button—Save).

The CFO hopes that one or more of the scenarios will point the way in the negotiations. But it is possible that none of them will show a profitable 2011, with debt less than $45 million. That would be a disaster for the company. After yet another four-year labor contract, the company not only would be losing money but also would be further in debt.

Thus, if none of the five scenarios is acceptable, you should use the input cells in a "what-if" manner. Change variables as you see fit to come up with an acceptable result. To do that, work just with the face of the spreadsheet—that is, do not add a new scenario to the Scenario Manager. There are some limitations to working this way: average ticket prices cannot increase or decrease more than ten cents per year, and Salaries and Benefits cannot increase or decrease by more than 2 percent. But the benefit is that you can find a reasonable way to make money and reduce debt if such a way exists. When you are finished, record the input variable values on a sheet of paper and call this scenario Last Ditch. If you are unable to come up with a good scenario, record the values of your best attempt.

Assignment 2B: Documenting Your Recommendation in a Memo

Open Word and write a brief memo to the company's CFO. Your goal is to offer negotiation guidance. Tell the CFO which of the five scenarios would yield profits and reduce debt in 2011. If none of the CFO's scenarios succeeded but you were able to find a Last Ditch scenario manually, tell the CFO about your results. Take a stand. In your opinion, which scenario offers the best negotiating position for the company? If you can see no way forward for the company, state that conclusion. Here are further instructions for writing your memo:

- Include a proper heading (TO/FROM/DATE/SUBJECT). You may use a Word memo template (Click **Office Button**, click **New**, click the **Memos** template in the **Templates** section, choose **Contemporary Memo**, then click **OK**. The first time you do this you may need to click **Download** to install the template).
- Do not provide background—the CFO is aware of the upcoming negotiations. Briefly state your analytical method and the results. Give the CFO your recommendation.
- Support your statements graphically as your instructor requires: (1) If you used the Scenario Manager, your instructor may want you to go back into Excel and copy the results of the Scenario Manager summary sheet to the Windows Clipboard. Then in Word, you copy this graphic into the memo. (Tutorial C discusses this procedure.) (2) Your instructor may want you to create a summary table in Word based on the results of the Scenario Manager summary sheet. The procedure for creating a table in Word is described next.

Enter a table into a Word document using the following procedure:

1. Click the cursor where you want the table to appear in the document.
2. In the **Insert** group, select the **Table** drop-down menu.
3. Select **Insert Table**.
4. Choose the number of rows and columns.
5. Click **OK**.

Your table should resemble the format shown in Figure 6-8.

Scenario	2011 Net Income	2011 Cash on Hand	2011 Debt Owed
No Change			
Traditional			
Hopeful			
Draconian			
Lower Price			
Last Ditch			

FIGURE 6-8   Format of table to insert in memo

In the table, include the Last Ditch scenario only if it was necessary in the analysis.

ASSIGNMENT 3: GIVING AN ORAL PRESENTATION

Your instructor may request that you present your analysis and recommendation in an oral presentation. If so, assume the CFO has accepted your analysis and recommendation. He has asked you to give a presentation explaining your recommendation to the company's senior management and the company's banker. Prepare to explain your analysis and recommendation to the group in ten minutes or less. Use any visual aids or handouts that you think are appropriate. Tutorial F has information on how to prepare and give an oral presentation.

DELIVERABLES

Assemble the following deliverables for your instructor:

1. Printout of your memo
2. Spreadsheet printouts
3. Disk, Zip disk, or jump drive, which should contain your Word memo file and Excel spreadsheet file. Do not provide a CD, as it would be read-only.

Staple the printouts together, placing the memo on top. If there is more than one .xlsx file on your disk, write your instructor a note stating the name of your model's .xlsx file.

THE TOLL ROAD LEASING DECISION

Decision Support Using Excel

PREVIEW

Your company operates toll roads. A Midwestern state offers to let your company take over and operate a 160-mile toll road for 30 years. In this case, you will use Excel to determine whether the operating agreement would be sufficiently profitable for your company.

PREPARATION

Before attempting this case, you should:

- Review spreadsheet concepts discussed in class and/or in your textbook.
- Complete any exercises that your instructor assigns.
- Complete any part of Tutorial C that your instructor assigns or refer to it as necessary.
- Review file-saving procedures for Windows programs, which are discussed in Tutorial C.
- Refer to Tutorials E and F as necessary.

BACKGROUND

You are a financial analyst for the Transportation Management Company (TMC). The company is in the toll road business, which works this way: A state in need of cash will let TMC take over and operate one of its toll roads. TMC gives the state a large amount of cash up front. TMC oversees the toll road for an agreed-upon number of years, collecting revenues and maintaining the road. At the end of the agreement, the state resumes operation of the road.

A large state in the middle of the United States currently needs cash. At first, the governor wanted to raise taxes, but the legislature refused to consider that option. So the governor contacted TMC, offering the state's 160-mile east-west toll road for 30 years in return for $2 billion dollars.

At this point, it looks as if the legislature is likely to approve the offer; so TMC has a decision to make. If the agreement looks sufficiently profitable, TMC will go forward; otherwise, there's no deal. The terms of the offer are set forth in the upcoming section, followed by a discussion of the toll road's operating requirements. After that, the objectives of TMC management are discussed.

Terms of the Offer

The state would still own the toll road. TMC would pay $2 billion at the end of 2007 to take control of the road from 2008 through 2037. At the end of 2037, the agreement would end and the state would resume as toll road operator.

TMC would keep all tolls it collects. The company also would keep all revenue collected at the toll road's rest stops, which sell gasoline and food. Tolls are charged per mile. Trucks are charged more than cars. TMC would not be able to raise tolls the first five years of the agreement, but could raise tolls each subsequent year within limits.

Two hundred and fifty state employees have been identified as toll road employees. They are involved in road maintenance and administrative tasks. Under the agreement, those employees would remain employed by the state, but their salaries would be paid by TMC. TMC would be allowed to make hiring and firing decisions. Maximum salary and benefits are set by the agreement, and any pay and benefits amounts above the maximum would be the obligation of the state.

TMC would borrow the $2 billion from banks in 2007 and give the funds to the state. The annual interest rate of the loan would be 3 percent. At first, that rate might seem low. However, because the toll road gets plenty of use, the bankers think that funds should be available for repayments. Because of the low risk, the bankers are able to accept a lower interest rate. TMC would repay as much loan principal as possible each year.

Under the $2 billion loan agreement, TMC would be required to keep at least $4 million on hand for operations. If there is a cash shortfall to begin any year, the banks stand ready to lend more money to TMC.

The state has a number of major road improvement projects planned for the upcoming five years. TMC would be required to finish those projects as planned, financing them out of toll road operations. One project is to install E-ZPass (and other related) booths at key spots along the 160-mile road. A second project involves performing a thorough surface upgrade in most of the cities through which the road passes. A third project involves upgrading some of the toll road's bridges. The company expects to spend $100 million in 2008 for such improvements. In addition, it has budgeted capital outlays for the next five years as follows: $75 million in 2009, $50 million in 2010, $40 million in 2011, and $25 million in 2012. It's difficult to predict the major outlays that will be needed after that five-year period, so the company assumes that it will spend $10 million a year through 2037 for capital projects. (A major improvement is referred to as a capital project to distinguish it from a maintenance project.)

Operating Requirements

Operating costs for the toll road will comprise (1) employee salaries and benefits, (2) fixed costs, and (3) road maintenance.

TMC engineers have developed a formula that lets them estimate road maintenance costs. One term in the formula is the price of a barrel of oil. TMC engineers believe that road maintenance costs are heavily influenced by the cost of oil, as many of the materials used in road maintenance are extracted from oil. Another term in the formula is the "cost adjustment factor." The cost adjustment factor varies with the price of oil. The relationship is shown in Figure 7-1.

Price of a Barrel of Oil	Cost Adjustment Factor
$ 50	8.00
65	7.50
70	7.00
75	6.50
80	6.00
85	5.50
90	5.00
95	4.50
100	4.25
110	4.10
120	3.90
130	3.75

FIGURE 7-1   Cost adjustment factors

Maintenance costs are expected to be a function of the price of a barrel of oil and the cost adjustment factor. The formula that TMC's engineers developed to describe the relationship between maintenance costs and the price of a barrel of oil is as follows: cost of maintenance per mile of road = (price of oil ^ 2) * cost adjustment factor. Example: Assume the price of a barrel of oil costs $100 this year. The maintenance cost per mile this year is estimated at ($100 ^ 2) * 4.25 = $42,500. That value represents the estimated cost if technology improvements were to play no role. But if technology can be expected to help reduce costs, the cost per mile would be 90 percent of $42,500.

TMC hopes that technology upgrades will enhance profitability by lowering the costs of operations, maintenance, and administration. Most of the managers at TMC believe that technology leads to improved productivity over time. Those managers also believe that improvements in technology result in fewer employees. For example, installing the E-ZPass system should translate into fewer toll booth collectors. However, a small group of managers do not think that technology improvements will be a significant factor in reducing costs; they believe the number of employees will be unaffected by upgrades in technology.

In addition to operating costs, TMC must consider various aspects of revenue collection. Revenue includes tolls collected and service stop concessions. Toll revenue is a function of mileage, the toll per mile, the traffic volume, and the type of vehicle. Following is a list of factors that must be considered when estimating toll revenue:

- Vehicle types are cars and trucks. Many more cars are on the road than trucks.
- The average trip length for a car is 50 miles, and the average trip length for a truck is 100 miles. Those averages are not expected to change.
- The length of the road, 160 miles, is not expected to change. The road could accommodate added lanes in some areas; but for the most part, the road will not change much in the coming years.
- The toll road is heavily used. Car and truck traffic is expected to grow at a rate of 1 percent per year in the next few years. Over time, however, that growth rate is expected to diminish. To be conservative, management assumes that the road will be maxed out eventually—in other words, there will be no growth in traffic volume in the last years of the agreement.
- Currently, cars are charged 10 cents a mile and trucks are charged 19 cents a mile. According to the terms of the agreement, TMC is not allowed to raise those rates in the first few years of its administration. After that, rates may be changed, but they will be subject to limits based on changes in the consumer price index (CPI).

Anyone who has traveled a toll road in the United States knows that the price of food and gasoline at service stops is very high. Thus, you would think that the concession revenue from service stops would be very profitable. However, on this state's toll roads, the average concession revenue per trip is surprisingly low. In fact, on this particular toll road, the concession revenue per trip is usually zero! Most travelers fill their gas tanks before (and/or after) they use the toll road; and they bring their own food, which they eat in their vehicles or at rest stop picnic tables. These days the average concession revenue per trip is $1.02. Under the agreement, concession prices can be increased, but only within limits set by the state. Also, if the concession revenues collected are above the average, the extra amount must be remitted to the state.

Management's Objectives

TMC has been operating toll roads around the country for many years. It's a very profitable business. TMC need not worry about building the roads—the state has already done that. The company simply pays money to the state, runs the road, then relinquishes the road back to the state. Because toll roads are heavily traveled in the United States, it usually is not difficult for TMC to make a profit.

The key for TMC is to structure the agreement so it will be sufficiently profitable. Doing so is complicated by the fact that the agreement calls for certain concessions that are good for the citizens but that do not favor TMC. For example: (1) TMC cannot raise toll rates for the first five years, and (2) concession revenue cannot exceed state-specified amounts. On the other hand, certain aspects of the agreement do favor TMC: (1) Pay and benefits increases are limited, (2) the company can hire and fire employees, (3) and the debt interest rate is reasonable.

Profit is sometimes called the return. The rate of return is the ratio of the profit to the amount invested. For multiyear projects, the rate of return is often stated using the time value of money. In this case, the rate can be referred to as the internal rate of return (IRR). The following example explains those various concepts.

Suppose a friend asks you to invest $20,000 in her beauty salon for three years. She says that she will pay you back in three installments: $7,000 after a year, $8,000 after two years, and $9,000 after three years. Thus, you have a $20,000 outflow now in return for future cash inflows of $7,000, $8,000, and $9,000. Is that a good deal for you?

Notice that your friend is willing to pay you $24,000 in the three years, which is $4,000 more than you invested. Therefore, you might say that you are making a 20 percent return on your $20,000 ($4000/$20,000 × 100). Does that mean the investment's IRR also is equal to 20 percent?

It does not! Why not? Because the value of a dollar paid back a year from today is not the same as the value of a dollar in your possession today. In other words, if you must wait a year to receive a dollar, the value of that dollar is less than the value of a dollar in today's terms. How much less? That depends on the interest rate at which you can invest your money. Assume you can invest your money at 10 percent interest. The value of a dollar a year from now is the amount of money that needs to be invested today at 10 percent so it will grow into a dollar a year from now.

Assume that amount of money is represented by x. Then x today + $.1x$ earned in the year = 1.00 received in a year. Another way to view that formula is as follows: $x * (1 + .1) = 1$. By algebraically solving for x, you get the following formula: $x = 1 / (1 +.1)$. Note that $1 / (1 + .1)$ equals .91. That .91 is called the *discount factor*. Thus, for the example involving an interest rate of 10 percent, the value of the dollar to be received a year from now is "discounted" by this factor and, therefore, would be worth 91 cents to you today. Note that 91 cents invested today at 10 percent grows to a dollar in a year—that is, $\$.91 * (1.1) = \$1.00$.

If you had to wait two years to get your dollar, the discount factor for an interest rate of 10 percent would be $1 / (1 + .1) ^ 2$, or .83, and the value today of the dollar received two years hence would be 83 cents. If you had to wait three years to get your dollar, the discount factor for an interest rate of 10 percent would be $1 / (1 + .1) ^ 3$, or .75, and the value of the dollar three years hence would be 75 cents.

Returning to the example of your friend's salon, suppose your goal is to earn a return of 10 percent on the investments you made in your friend's salon. Figure 7-2 shows the results when you apply the 10 percent discount factors to the cash inflows your friend has promised you.

Year	Annual Repayments	Discount Factor	Inflows in Todayís Dollars
1	$ 7,000	$1 / (1 + .1) ^ 1 = .91$	$ 6,370
2	8,000	$1 / (1 + .1) ^ 2 = .83$	6,640
3	9,000	$1 / (1 + .1) ^ 3 = .75$	6,750
Total	24,000		19,760

FIGURE 7-2   Repayments in todayís dollars at 10 percent

You will notice that the inflows stated in today's dollars total less than the $20,000 you are asked to contribute now. That means you are not getting your required 10 percent rate of return over the three years. In other words, the investment's IRR is less than 10 percent—and, therefore, not sufficient.

Notice that there must be an interest rate at which the three discounted inflows equal the $20,000 contribution. So another way to look at this sort of investment is to find out its actual IRR and then compare that to the rate you require as an investor. If the actual IRR is greater than your required rate, the investment is a good one for you. If the actual IRR is less than your required rate, the investment is not a good one for you. There are a number of ways to compute IRR, either by using a mathematical formula or by using tables to approximate the IRR. Fortunately, Excel has an IRR function that performs this calculation.

Returning again to the salon investment example, the IRR that is needed to make the future discounted inflows total the $20,000 you invested is 9.28 percent, as shown in Figure 7-3:

Year	Annual Repayments	Discount Factor	Inflows in Todayís Dollars
1	$ 7,000	$1 / (1 + .0928) ^ 1 = .91508$	$ 6,405
2	8,000	$1 / (1 + .0928) ^ 2 = .83737$	6,699
3	9,000	$1 / (1 + .0928) ^ 3 = .76626$	6,896
Total	24,000		20,000

FIGURE 7-3   Repayments in todayís dollars at 9.28 percent

Since you required a 10 percent minimum annual rate of return on your investment, your friend's proposal would not be acceptable to you, because a 9.28 percent IRR is less than the minimum. But if you had required only a 9 percent rate of return, the proposal would be acceptable to you.

TMC management uses 10 percent as the corporate IRR goal. TMC will consider accepting proposed investments that look as if they will earn 10 percent or more. TMC will not consider investments that look as if they will earn less than 10 percent.

The company needs a spreadsheet that shows a projected income statement and cash flow statement, given assumptions about the effects of technology, the rate of inflation, and the price of oil. You have been called in to make the projection for 2007–2037. Certain likely scenarios are under consideration, and your spreadsheet will model them in the Scenario Manager. Which of the scenarios will be sufficiently profitable for TMC? Should TMC accept the state's offer?

ASSIGNMENT 1: CREATING A SPREADSHEET FOR DECISION SUPPORT

In this assignment, you will produce a spreadsheet that models the business decision. Then in Assignment 2, you will write a memo to the company's CFO that explains your findings and recommendation. In addition, in Assignment 3, you will be asked to prepare an oral presentation of your analysis.

First, you will create a spreadsheet that models the toll road offer. The model is an analysis of the offer covering 2007–2037. You will be given some hints on how to set up each section before you enter cell formulas. Your spreadsheet should have the following sections:

- CONSTANTS
- INPUTS
- SUMMARY OF KEY RESULTS
- CALCULATIONS
- INCOME STATEMENT AND CASH FLOW STATEMENT
- DEBT OWED
- NET INCOME OVER LIFE OF PROJECT
- IRR CALCULATION
- VLOOKUP TABLE

A discussion of each of those sections follows. The spreadsheet skeleton is available to you, so you do not need to type it in. To access the spreadsheet, go to your Data files. Select Case 7 and then select TOLLROAD.xlsx.

CONSTANTS Section

Your spreadsheet should have the constants shown in Figure 7-4. Note that only some of the values for the 30-year period of the agreement are shown. An explanation of the line items follows the figure.

	A	B	C	D	E
1	TOLL ROAD LEASE DECISION				
2					
3	CONSTANTS	2007	2008	2009	2010
4	TAX RATE	NA	0.20	0.20	0.20
5	CASH NEEDED TO START NEXT YEAR	NA	4000000	4000000	4000000
6	FIXED OPERATING COSTS	NA	10000000	10010000	10020000
7	INTEREST RATE ON DEBT	NA	0.03	0.03	0.03
8	EXPECTED CAPITAL PROJECTS	NA	100000000	75000000	50000000
9	EXPECTED TRAFFIC GROWTH	NA	0.01	0.01	0.01
10	AVERAGE TRIP LENGTH -- CARS	NA	50	50	50
11	AVERAGE TRIP LENGTH -- TRUCKS	NA	100	100	100
12	MAXIMUM AVERAGE SALARY & BENEFITS	NA	62500	62600	62700
13	MAXIMUM CONCESSION REVENUE PER TRIP	NA	1.070	1.100	1.130

FIGURE 7-4 CONSTANTS section

- TAX RATE: The expected tax rates on income before taxes for 2008–2037 are given.
- CASH NEEDED TO START NEXT YEAR: TMC's bankers say that the company needs to start each year with a certain amount of cash. Each year that amount is $4 million.
- FIXED OPERATING COSTS: Fixed costs are operating costs other than maintenance and employment costs. The amount is expected to be $10 million in 2008, increasing somewhat each year thereafter.
- INTEREST RATE ON DEBT: This is expected to be 3 percent each year.

- EXPECTED CAPITAL PROJECTS: Major construction projects are expected to be costly the first few years, but they will level out at $10 million a year.
- EXPECTED TRAFFIC GROWTH: The number of car and truck trips is expected to increase 1 percent per year for 2008–2011. Traffic growth is expected to be a half a percent per year from 2012–2015. Traffic growth is expected to decline with time, and it should be zero by the end of the agreement.
- AVERAGE TRIP LENGTH—CARS: Some trips by auto are for the entire 160-mile length of the road. Some trips are for just a few miles within a city. For all of the years of the agreement, the average length of car trips is expected to be 50 miles.
- AVERAGE TRIP LENGTH—TRUCKS: For all of the years of the agreement, the average length of truck trips is expected to be 100 miles.
- MAXIMUM AVERAGE SALARY & BENEFITS: Toll road workers remain state employees, but they work for TMC during the agreement. What if the state employee union negotiates a substantial pay increase for all state workers? Is TMC required to pay for that increase in salaries (or benefits)? Under the agreement, the company would not be obligated; instead, the state would cover the difference. Therefore, the yearly average salary and benefits ceilings are given here.
- MAXIMUM CONCESSION REVENUE PER TRIP: Most toll road drivers do not buy gas or food at the toll road rest stop, but some travelers do. What is to prevent TMC from raising the prices of rest stop concessions to outrageous levels? The state imposes a limit on concession prices; therefore, the average amount allowed to be charged per trip is given. (What if, on average, people spend more than that amount? TMC would have to rebate the difference to the state.)

INPUTS Section

Your spreadsheet should have the inputs shown in Figure 7-5. *NA* (not applicable) designates a cell that should not be used to enter inputs; *NA* also indicates that the cell address should not be referenced in any cell's formula. An explanation of the line items follows the figure.

	A	B	C	D	E
15	INPUTS	ALL YRS	2008	2009	2010
16	TECHNOLOGY HELPS (Y/N)		NA	NA	NA
17	AVG CONSUMER PRICE INDEX INCREASE (.XX)		NA	NA	NA
18	PRICE OF BARREL OF OIL	NA			

FIGURE 7-5 INPUTS section

- TECHNOLOGY HELPS (Y/N): Will technology advances be sufficient to reduce the number of employees during the life of the agreement? If the user thinks the answer is yes, he or she enters a *Y*. If the user thinks the answer is no, he or she enters an *N*. The value entered applies to all 30 years.
- AVG CONSUMER PRICE INDEX INCREASE (.XX): The CPI is a measure of inflation. A rise in the index of .01 corresponds to a 1 percent increase in inflation. If the user thinks the average inflation increase each year will be 1 percent, he or she enters .01. The value entered applies to all 30 years.
- PRICE OF BARREL OF OIL: The price of oil will affect maintenance costs each year. The price expected each year is entered here. Assume the user thinks that the price will be $80 in 2008, $90 in 2009, $100 in 2010, and then $110 until the end of the agreement. In that case, he or she would enter 80, 90, 100, and then twenty-seven 110s.

Your instructor may tell you to apply Conditional Formatting to one or more of the input cells so that out-of-bounds values are highlighted in some way. (For example, the entry could show up in red type or in bold-faced type.) If so, your instructor may provide a tutorial on Conditional Formatting or ask you to refer to Excel Help.

SUMMARY OF KEY RESULTS Section

Your spreadsheet should show these results: (1) net income after taxes for the entire agreement period, (2) cash on hand at the end of the agreement, (3) debt owed at the end of the agreement, and (4) the investment's IRR. The format is shown in Figure 7-6. These values are all computed elsewhere in the spreadsheet and should be echoed here.

	A	B	C	D	E
20	SUMMARY OF KEY RESULTS	ALL YRS			
21	NET INCOME AFTER TAXES -- ALL 30 YEARS		NA	NA	NA
22	END-OF-THE-YEAR CASH ON HAND -- 2037		NA	NA	NA
23	END-OF-THE-YEAR DEBT OWED -- 2037		NA	NA	NA
24	INTERNAL RATE OF RETURN		NA	NA	NA

FIGURE 7-6 SUMMARY OF KEY RESULTS section

CALCULATIONS Section

You should calculate various intermediate results that will be used in the income statement and cash flow statement that follows. Those calculations, shown in Figure 7-7, are based on inputs and on 2007 values. When called for, use absolute addressing properly. An explanation of the line items follows the figure.

	A	B	C	D	E
26	CALCULATIONS	2007	2008	2009	2010
27	NUMBER OF EMPLOYEES	250			
28	AVERAGE SALARY & BENEFITS	60000			
29	AVERAGE TOLL/MILE -- CARS	0.100			
30	AVERAGE TOLL/MILE -- TRUCKS	0.190			
31	NUMBER OF TRIPS -- CARS	24000000			
32	NUMBER OF TRIPS -- TRUCKS	6000000			
33	MAINTENANCE COST PER MILE	31000			
34	CONCESSION REVENUE PER TRIP	1.020			

FIGURE 7-7 CALCULATIONS section

- NUMBER OF EMPLOYEES: If technology will help lower costs, each year the number of toll road employees will be one person less than the prior year. Otherwise, the number of employees in a year will be the same as the prior year.
- AVERAGE SALARY & BENEFITS: Average employee pay (salary and benefits) in a year is a function of the prior year's salary and the increase in CPI. Example: The average in 2007 is 60,000, and a 1 percent increase in CPI is expected. Therefore, the 2008 salary would be $60,600. The average in 2009 would be 1 percent more than 2008's $60,600, the average in 2010 would be 1 percent more than 2009's $61,206, etc. You should note, however, that the average salary in a year cannot exceed the maximum allowed in the year; that is, the maximum in a year is a constant. To determine the lesser of the two amounts—the calculated average pay and the maximum average pay allowed—you can use the MIN() function or an IF() statement.
- AVERAGE TOLL/MILE—CARS: The average toll per mile is fixed until 2013. Then the average toll per mile in a year is a function of the prior year's average toll per mile and the increase in CPI. According to the agreement, the company can increase the average toll per mile by the amount of the CPI increase in a year. Example: The average toll is 10 cents in 2012, and the 2013 increase in CPI is expected to be 1 percent. Therefore, the average toll in 2013 can be 10 cents plus 1 percent of 10 cents. The average toll in 2014 would be 2013's average toll, plus 1 percent of that. If, however, the CPI increase exceeds 2 percent, there is a limit to what the average toll increase can be. In that case, the toll can increase by only 2 percent plus one-half of the CPI increase in excess of 2 percent. For example, if the CPI increase is 4 percent, the toll can increase by the following value: 2% + (.5 * (4% – 2%)). That results in a toll increase of 3 percent.
- AVERAGE TOLL/MILE—TRUCKS: The logic that applied to the calculation of the average toll per mile for cars applies here.

- NUMBER OF TRIPS—CARS: The number of trips taken by car in a year is a function of the number of trips taken the prior year and the year's traffic growth factor, which is a constant from Figure 7-4. Example: In a given year, traffic growth is expected to be 1 percent; and in the prior year, 25 million car trips were taken. Therefore, this year the number of trips taken will be 25 million plus 1 percent of 25 million.
- NUMBER OF TRIPS—TRUCKS: The logic that applied to car traffic applies here.
- MAINTENANCE COST PER MILE: This value is a function of the price of a barrel of oil that is expected for the year and the related cost increase factor. The cost increase factor is obtained by a VLOOKUP table at the bottom of the spreadsheet. Additionally, when technology is expected to help operations (an input value), the cost is 90 percent of the calculated amount.
- CONCESSION REVENUE PER TRIP: The average concession revenue per trip is a function of the prior year's average value and the increase in CPI. According to the agreement, the company can increase the concession revenue by the amount of the CPI increase in a year. Example: The CPI increase in a given year is expected to be 10 percent. The average revenue per trip in the prior year was $1.00. In the current year, the average concession revenue per trip could increase to $1.10. You should note, however, that the average value in a year cannot exceed the maximum allowed in the year; that is, the maximum in a year is a constant. To determine the lesser of the two amounts—the calculated concession revenue per trip and the maximum concession revenue per trip allowed—you can use the MIN() function or an IF() statement.

INCOME STATEMENT AND CASH FLOW STATEMENT Section

The forecast for net income and cash flow starts with the cash on hand at the beginning of the year. That is followed by the income statement, after which comes the calculation of cash on hand at year's end. For readability, format cells in this section for zero decimals. Your spreadsheet should look like those shown in Figure 7-8 and Figure 7-9. A discussion of line items follows each figure.

	A	B	C	D	E
37	INCOME STATEMENT AND CASH FLOW STATEMENT	2007	2008	2009	2010
38	BEGINNING-OF-THE-YEAR CASH ON HAND	NA			
39					
40	TOLL REVENUE – CARS	NA			
41	TOLL REVENUE – TRUCKS	NA			
42	CONCESSION REVENUE	NA			
43	TOTAL REVENUE	NA			
44	SALARY AND BENEFITS COSTS	NA			
45	MAINTENANCE COSTS	NA			
46	FIXED COSTS	NA			
47	TOTAL COSTS	NA			
48	INCOME BEFORE INTEREST AND TAXES	NA			
49	INTEREST EXPENSE	NA			
50	INCOME BEFORE TAXES	NA			
51	INCOME TAX EXPENSE	NA			
52	NET INCOME AFTER TAXES	NA			

FIGURE 7-8   INCOME STATEMENT AND CASH FLOW STATEMENT section

- BEGINNING-OF-THE-YEAR CASH ON HAND: This is the cash on hand at the end of the prior year.
- TOLL REVENUE—CARS: This amount is a function of the number of car trips, the average length of a car trip, and the average toll per mile for cars. These values are either constants or calculations.
- TOLL REVENUE—TRUCKS: This amount is a function of the number of truck trips, the average length of a truck trip, and the average toll per mile for trucks. These values are either constants or calculations.
- CONCESSION REVENUE: This amount is a function of concession revenue per trip and the number of car and truck trips. These are all calculated values.
- TOTAL REVENUE: This amount is the total of toll and concession revenue.
- SALARY AND BENEFITS COSTS: The amount is a function of the number of toll road employees and the average salary and benefits cost per employee for the year. These are all calculated values.

- MAINTENANCE COSTS: This amount is a function of the maintenance cost per mile (a calculation) and the length of the toll road, 160 miles.
- FIXED COSTS: This amount is a constant, which can be echoed here.
- TOTAL COSTS: This amount is the sum of salary and benefits costs, maintenance costs, and fixed costs.
- INCOME BEFORE INTEREST AND TAXES: This amount is the difference between total revenue and total costs.
- INTEREST EXPENSE: This amount is a function of the debt owed at the beginning of the year and the interest rate on debt, a constant. The interest rate applies to the $2 billion borrowed to finance the agreement and to any borrowings necessitated by cash shortfalls in a year.
- INCOME BEFORE TAXES: This amount is equal to income before interest and taxes, minus interest expense.
- INCOME TAX EXPENSE: This amount is zero when income before taxes is zero or negative. Otherwise, income tax expense is a function of the year's tax rate and the income before taxes.
- NET INCOME AFTER TAXES: This is the difference between income before taxes and income tax expense.

Figure 7-9 shows the next section of this statement, a computation of cash on hand at the end of the year. A discussion of the line items for this year-end cash calculation follows the figure. Note that in Figure 7-9, column B corresponds to 2007; column C, to 2008; etc. Also, 2007 values are NA, except for END-OF-THE-YEAR CASH ON HAND, which is $4,000,000.

	A	B	C	D	E
54	CAPITAL PROJECT EXPENDITURES	NA			
55					
56	NET CASH POSITION (NCP) BEFORE BORROWING AND REPAYMENT OF DEBT (BEG OF YR CASH + NET INCOME - CAPITAL PROJECTS)	NA			
57	ADD:BORROWING FROM BANK	NA			
58	LESS: REPAYMENT TO BANK	NA			
59	EQUALS: END-OF-YEAR CASH ON HAND	4000000			

FIGURE 7-9 END-OF-THE-YEAR CASH ON HAND section

- CAPITAL PROJECT EXPENDITURES: This cash outflow is for major road projects in a given year. The amount is a constant that can be echoed here.
- NET CASH POSITION (NCP): The NCP at the end of a year equals cash at the beginning of the year plus the year's net income after taxes, less the outflows for capital projects.
- ADD: BORROWING FROM BANK: Assume that the company's bankers will lend the company enough money at year end to get to the minimum cash needed to start the next year. If the NCP is less than this minimum, the company must borrow enough money to get to the minimum. Borrowing increases cash on hand, of course.
- LESS: REPAYMENT TO BANK: If the NCP is more than the minimum cash at the end of a year and debt is owed, the company must pay off as much debt as possible (but not so much that the cash falls below the minimum cash required to start the next year). Repayments reduce cash on hand, of course.
- EQUALS: END-OF-YEAR CASH ON HAND: This equals the NCP plus any borrowings, less any repayments.

DEBT OWED Section

The body of your spreadsheet continues with a calculation of debt owed at year end, as shown in Figure 7-10. An explanation of line items follows the figure. The values for 2007 are NA, except that the company owes $2,000,000,000 at the end of that year.

- BEGINNING-OF-THE-YEAR DEBT OWED: Cash owed at the beginning of a year equals cash owed at the end of the prior year.
- ADD: BORROWING FROM BANK: This amount has been calculated elsewhere and can be echoed in this section. Borrowings increase debt owed.

	A	B	C	D	E
61	**DEBT OWED**	**2007**	**2008**	**2009**	**2010**
62	BEGINNING-OF-YEAR DEBT OWED	NA			
63	ADD: BORROWING FROM BANK	NA			
64	LESS: REPAYMENT TO BANK	NA			
65	EQUALS: END-OF-THE-YEAR DEBT OWED	2000000000			

FIGURE 7-10   DEBT OWED section

- LESS: REPAYMENT TO BANK: This amount has been calculated elsewhere and can be echoed in this section. Repayments reduce debt owed.
- EQUALS: END-OF-THE-YEAR DEBT OWED: This equals the amount owed at the beginning of a year plus borrowings in the year, less repayments in the year.

NET INCOME—LIFE OF PROJECT Section

Management wants to know the total net income after taxes that will have been earned in the 30 years of the agreement. The sum should be recorded in a cell below the **DEBT OWED** section, as shown in Figure 7-11. *Note:* This net income for the life of the project should be echoed in the Summary of Key Results.

	A	B	C	D	E
67	**NET INCOME -- LIFE OF PROJECT**		NA	NA	NA

FIGURE 7-11   NET INCOME—LIFE OF PROJECT section

INTERNAL RATE OF RETURN (IRR) Section

The IRR for the 30 years is computed here. The computation is done in two steps, as shown in Figure 7-12.

	A	B	C	D	E
69	**INTERNAL RATE OF RETURN**	**2007**	**2008**	**2009**	**2010**
70	**CASH FLOW IN YEAR**	-2000000000			
71	**IRR FOR PROJECT**				

FIGURE 7-12   INTERNAL RATE OF RETURN (IRR) section

- CASH FLOW IN YEAR: This is the cash flow from operating the road in a year; it is equal to the net income after taxes, less capital project expenditures. If the net income is high enough, the cash flow value will be positive in a year. Note that in 2007, there is only one cash flow—the $2 billion outflow required to enter the agreement with the state. Note also that there will be 31 cash flow values in your spreadsheet: the first-year outflow and 30 net cash flows for each operating year.
- IRR FOR PROJECT: The IRR is the rate that equalizes the first-year outflow with the discounted values for the 30 operating years. Use the IRR() function to compute this value. The syntax for IRR() is =IRR(Range for cash flows, interest rate guess). Here, the range of flows is the 2007–2037 cash flows. You can use –10% as the guess. Note the minus sign and the percent sign in the guess. The IRR value should be echoed in the Summary of Key Results.

VLOOKUP TABLE Section

You need a VLOOKUP table to get the correct maintenance cost adjustment factor. You should place the table, shown in Figure 7-13, somewhere below the IRR section.

B	C	D
	PRICE OF BARREL OF OIL	COST ADJUSTMENT FACTOR
	50	8.000
	65	7.500
	70	7.000
	75	6.500
	80	6.000
	85	5.500
	90	5.000
	95	4.500
	100	4.250
	110	4.100
	120	3.900
	130	3.750

FIGURE 7-13 VLOOKUP TABLE section

ASSIGNMENT 2: USING THE SPREADSHEET FOR DECISION SUPPORT

You will now complete the case by (1) using the spreadsheet to gather the data needed to assess the impact of technology, CPI increases, and fluctuations in the price of oil and (2) documenting your findings and recommendation in a memo.

TMC's CFO wants to know what combinations of the key variables will lead to an IRR of at least 10 percent. The CFO has six scenarios in mind and wants to know the result of each. That knowledge will help management decide if entering into a contract with the state is a good idea.

Assignment 2A: Using the Spreadsheet to Gather Data

You have built the spreadsheet to model the business situation. The CFO wants to know values for the following factors for each of the six scenarios he believes may occur: profitability in all years, cash on hand at the end of the project, debt owed at the end of the project, and the project's IRR. In all scenarios, technology is assumed to be a positive factor. The scenarios are described in the following notations:

- LOWINF-HIGHOIL: Technology will help. The increase in the CPI is expected to be just 1 percent a year, which would be considered low inflation. Oil per barrel will be $80 in 2008, $90 in 2009, $100 in 2010, and $100 in the years thereafter.
- MIDINF-HIGHOIL: Technology will help. The increase in the CPI is expected to be 4 percent a year, which would be considered a midrange inflation level. Oil per barrel will be $80 in 2008, $90 in 2009, $100 in 2010, and $100 in the years thereafter.
- HIGHINF-HIGHOIL: Technology will help. The increase in the CPI is expected to be 6 percent a year, which would be considered a high inflation level. Oil per barrel will be $80 in 2008, $90 in 2009, $100 in 2010, and $100 in the years thereafter.
- LOWINF_LOWOIL: Technology will help. The increase in the CPI is expected to be just 1 percent a year. Oil per barrel will be $80 in 2008, $90 in 2009, $100 in 2010, $100 in 2011, $90 in 2012, $80 in 2013, $70 in 2014, and $70 in the years thereafter.
- MIDINF-LOWOIL: Technology will help. The increase in the CPI is expected to be 4 percent a year. Oil per barrel will be $80 in 2008, $90 in 2009, $100 in 2010, $100 in 2011, $90 in 2012, $80 in 2013, $70 in 2014, and $70 in the years thereafter.
- HIGHINF-LOWOIL: Technology will help. The increase in the CPI is expected to be 6 percent a year. Oil per barrel will be $80 in 2008, $90 in 2009, $100 in 2010, $100 in 2011, $90 in 2012, $80 in 2013, $70 in 2014, and $70 in the years thereafter.

The financial results in each scenario will act as a negotiating guide to management. Using Scenario Manager, you should run "what-if" scenarios with the six sets of input values. See Tutorial C for the Scenario Manager procedures. Set up the six scenarios. The changing cells are the cells used to input the technology variable, the CPI increase, and the oil prices. (*Note:* In the Scenario Manager, you can enter noncontiguous cell ranges as follows: C20..F20, C21, C22. Cell addresses cited here are arbitrary). The output cells are the

Summary of Key Results cells. Run the Scenario Manager to gather the data in a report. When you are finished, print the spreadsheet with inputs for any one of the scenarios and print the Scenario Manager Summary Sheet. Then save the spreadsheet file one last time (Office Button—Save).

The CFO wants to know which scenarios, if any, lead to a 10 percent or greater IRR. If they all do, the state's offer should probably be accepted. If only some do, then management must think more deeply about committing to the project.

In addition, the CFO wants your opinion about how sensitive the results are to changes in each of the three key variables. Here are some guidelines for how to analyze this sensitivity:

- Technology: Work with scenario values on the face of the spreadsheet; in other words, you should not add a new scenario to the Scenario Manager. Enter one of the six scenario input values. Change the Technology variable to N. How much does the project's net income and the IRR change? Do this for a few of the scenarios to get an idea of how sensitive the IRR is to the Technology variable. Note your conclusions as a comment in a blank part of the spreadsheet or write your notes on a sheet of paper.
- CPI: The sensitivity of the CPI variable may be apparent from the Scenario Manager summary report. If not, employ the same procedure outlined for Technology. You may want to try values greater than 6 percent to see the effect. Note your conclusions as a comment in a blank part of the spreadsheet or write your notes on a sheet of paper. The CFO thinks that inflation greater than 3 percent is possible, but not likely, in the United States for 30 years.
- Price of oil: Again, the sensitivity may be clear from the Scenario Manager summary report. If not, employ the same procedure outlined for Technology. You may want to use different values or a different sequence of values. Note your conclusions as a comment in a blank part of the spreadsheet or write your notes on a sheet of paper.

Assignment 2B: Documenting Your Recommendation in a Memo

Open Word and write a memo to the company's CFO. Your goal is to offer negotiation guidance. You can organize your memo into three parts: analysis results, negotiating guidance, and recommendation. Each part is detailed below:

- Analysis Results: Tell the CFO which of the six scenarios would yield an IRR greater than 10 percent. Support your findings graphically.
- Negotiating Guidance: Tell the CFO which variables seem to have a significant impact on the results and which do not. For the sensitivity analysis, cite some financial results to justify your findings. Support them graphically with a table. The CFO should be able to understand from your findings whether a variable is essential for success in some way. For example, is it necessary that TMC raise toll rates substantially for the project to earn an acceptable IRR?
- Recommendation: Take a stand. In your opinion, which of the scenarios offers the best negotiating position for the company? If you can see no way forward for the company, state that conclusion.

Here are further instructions for writing your memo:

- Include a proper heading (TO/FROM/DATE/SUBJECT). You may use a Word memo template (Click **Office Button**, click **New**, click the **Memos** template in the **Templates** section, choose **Contemporary Memo**, then click **OK**. The first time you do this you may need to click Download to install the template).
- Organize the body into three sections; use headings to set these sections apart from each other.
- Do not provide background—the CFO is aware of the upcoming negotiations.
- Support your statements graphically as your instructor requires: (1) Your instructor may want you to go back into Excel and copy the results of the Scenario Manager summary sheet to the Windows Clipboard. Then in Word, you copy this graphic to the memo. (Tutorial C discusses this procedure.) (2) Your instructor may want you to create a summary table in Word based on the results of the Scenario Manager summary sheet. The procedure for creating a table in Word is described next.

Enter a table into a Word document using the following procedure:

1. Click the cursor where you want the table to appear in the document.
2. In the **Insert** group, select the **Table** drop-down menu.
3. Select **Insert Table**.
4. Choose the number of rows and columns.
5. Click **OK**.

Your table should resemble the format shown in Figure 7-14.

Scenario	Net Income, All Years	2037 Cash on Hand	2037 Debt Owed	IRR
Low Inflation, High Oil				
Mid Inflation, High Oil				
High Inflation, High Oil				
Low Inflation, Low Oil				
Mid Inflation, Low Oil				
High Inflation, Low Oil				

FIGURE 7-14   Format of table to insert in memo

You should use a table to document the results of your sensitivity analysis. Your table should resemble the format shown in Figure 7-15.

Variable	Impact on IRR	Justification
Technology helps?		
CPI change		
Price of barrel of oil		

FIGURE 7-15   Format of sensitivity analysis table to insert in memo

In the table, your justification should be one or two sentences that summarize the results of the what-if scenario analysis the CFO requested.

ASSIGNMENT 3: GIVING AN ORAL PRESENTATION

Your instructor may request that you present your analysis and recommendation in an oral presentation. If so, assume the CFO has accepted your analysis and recommendation. He has asked you to give a presentation explaining your recommendation to the company's senior management and the company's bankers. Prepare to explain your analysis and recommendation to the group in ten minutes or less. Use any visual aids or handouts that you think are appropriate. Tutorial E has guidance on how to prepare and give an oral presentation.

DELIVERABLES

Assemble the following deliverables for your instructor:

1. Printout of your memo
2. .xlsx spreadsheet printouts
3. Disk, Zip disk, or jump drive that contains your memo and spreadsheet files. Do not provide a CD, as it would be read-only.

Staple the printouts together, placing the memo on top.

PART 3

DECISION SUPPORT CASES
USING THE EXCEL SOLVER

BUILDING A DECISION SUPPORT SYSTEM USING THE EXCEL SOLVER

Decision Support Systems (DSS) help people make decisions. (The nature of DSS programs is discussed in Tutorial C.) Tutorial D teaches you how to use the Solver, one of the Excel built-in decision support tools.

For some business problems, decision makers want to know the best, or optimal, solution. Usually that means maximizing a variable (for example, net income) or minimizing another variable (for example, total costs). This optimization is subject to constraints, which are rules that must be observed when solving a problem. The Solver computes answers to such optimization problems.

This tutorial has four sections:

1. **Using the Excel Solver** In this section, you'll learn how to use the Solver in decision making. As an example, you use the Solver to create a production schedule for a sporting goods company. This schedule is called the Base Case.
2. **Extending the Example** In this section, you'll test what you've learned about using the Solver as you modify the sporting goods company's production schedule. This is called the Extension Case.
3. **Using the Solver on a New Problem** In this section, you'll use the Solver on a new problem.
4. **Troubleshooting the Solver** In this section, you'll learn how to overcome problems you might encounter when using the Solver.

> **NOTE**
>
> Tutorial C offers some guidance on basic Excel concepts, such as formatting cells and using functions such as =IF(). Refer to Tutorial C for a review of those topics.

USING THE EXCEL SOLVER

Suppose a company must set a production schedule for its various products, each of which has a different profit margin (selling price, less costs). At first, you might assume the company will maximize production of all profitable products to maximize net income. However, a company typically cannot make and sell an unlimited number of its products because of constraints.

One constraint affecting production is the shared resource problem. For example, several products in a manufacturer's line might require the same raw materials, which are in limited supply. Similarly, the manufacturer might require the same machines to make several of its products. In addition, there also might be a limited pool of skilled workers available to make the products.

In addition to production constraints, sometimes management's policies impose constraints. For example, management might decide that the company must have a broader product line. As a consequence, a certain production quota for several products must be met, regardless of profit margins.

Thus, management must find a production schedule that will maximize profit given the constraints. Optimization programs like the Solver look at each combination of products, one after the other, ranking each combination by profitability. Then the program reports the most profitable combination.

To use the Solver, you'll set up a model of the problem, including the factors that can vary, the constraints on how much they can vary, and the goal—that is, the value you are trying to maximize (usually net income) or minimize (usually total costs). The Solver then computes the best solution.

Setting Up a Spreadsheet Skeleton

Suppose your company makes two sporting goods products—basketballs and footballs. Assume you will sell all of the balls you produce. To maximize net income, you want to know how many of each kind of ball to make in the coming year.

Making each kind of ball requires a certain (and different) number of hours, and each ball has a different raw materials cost. Because you have only a limited number of workers and machines, you can devote a maximum of 40,000 hours to production, which is a shared resource. You do not want that resource to be idle, however. Downtime should be no more than 1,000 hours in a year, so machines should be used for at least 39,000 hours.

Marketing executives say that you cannot make more than 60,000 basketballs and cannot make fewer than 30,000. Furthermore, they say that you must make at least 20,000 footballs but not more than 40,000. Marketing also says that the ratio of basketballs to footballs produced should be between 1.5 and 1.7—that is, more basketballs than footballs, but within limits.

What would be the best production plan? This problem has been set up in the Solver. The spreadsheet sections are discussed in the pages that follow.

AT THE KEYBOARD

Start out by saving the blank spreadsheet as SPORTS1.xls. Then enter the skeleton and formulas as they are discussed.

CHANGING CELLS Section

The **CHANGING CELLS** section contains the variables the Solver is allowed to change while it looks for the solution to the problem. Figure D-1 shows the skeleton of this spreadsheet section and the values you should enter. An analysis of the line items follows the figure.

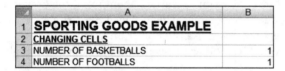

	A	B
1	**SPORTING GOODS EXAMPLE**	
2	CHANGING CELLS	
3	NUMBER OF BASKETBALLS	1
4	NUMBER OF FOOTBALLS	1

FIGURE D-1 CHANGING CELLS section

- The changing cells are for the number of basketballs and footballs to be made and sold. The changing cells are like input cells, except Solver (not you) plays "what if" with the values, trying to maximize or minimize some value (in this case, maximize net income).
- Some number should be put in the changing cells each time before the Solver is run. It is customary to put the number 1 in the changing cells (as shown). Solver will change these values when the program is run.

CONSTANTS Section

Your spreadsheet also should have a section for values that will not change. Figure D-2 shows a skeleton of the **CONSTANTS** section and the values you should enter. A discussion of the line items follows the figure.

> **NOTE**
>
> You should format cells in the constants range to two decimal places.

	A	B
6	**CONSTANTS**	
7	BASKETBALL SELLING PRICE	14.00
8	FOOTBALL SELLING PRICE	11.00
9	TAX RATE	0.28
10	NUMBER OF HOURS TO MAKE A BASKETBALL	0.50
11	NUMBER OF HOURS TO MAKE A FOOTBALL	0.30
12	COST OF LABOR -- 1 MACHINE HOUR	10.00
13	COST OF MATERIALS -- 1 BASKETBALL	2.00
14	COST OF MATERIALS -- 1 FOOTBALL	1.25

FIGURE D-2 CONSTANTS Section

- The SELLING PRICE for a single basketball and a single football is shown.
- The TAX RATE is the rate applied to income before taxes to compute income tax expense.
- The NUMBER OF MACHINE HOURS needed to make a basketball and a football is shown. Note that a ball-making machine can produce two basketballs in an hour.
- COST OF LABOR: A ball is made by a worker using a ball-making machine. A worker is paid $10 for each hour he or she works at a machine.
- COST OF MATERIALS: The costs of raw materials for a basketball and football are shown.

Notice that the profit margins (selling price, less costs of labor and materials) for the two products are not the same. They have different selling prices and different inputs (raw materials and hours to make), and the inputs have different costs per unit. Also note that you cannot tell from the data how many hours of the shared resource (machine hours) will be devoted to basketballs and how many will be devoted to footballs, because you don't know in advance how many basketballs and footballs will be made.

Calculations Section

In the **CALCULATIONS** section, you will calculate intermediate results that (1) will be used in the spreadsheet body and/or (2) will be used as constraints. Before entering formulas, format the Calculations range cells for two decimal places (cell formatting directions were given in Tutorial C). Figure D-3 shows the skeleton and formulas you should enter. A discussion of the cell formulas follows the figure.

N O T E

Cell widths are changed here merely to show the formulas—you need not change the width.

	A	B
16	**CALCULATIONS**	
17	RATIO OF BASKETBALLS TO FOOTBALLS	=B3/B4
18	TOTAL BASKETBALL HOURS USED	=B3*B10
19	TOTAL FOOTBALL HOURS USED	=B4*B11
20	TOTAL MACHINE HOURS USED (BB + FB)	=B18+B19

FIGURE D-3 CALCULATIONS section cell formulas

- RATIO OF BASKETBALLS TO FOOTBALLS: This number (cell B17) will be needed in a constraint.
- TOTAL BASKETBALL HOURS USED: The number of machine hours needed to make all basketballs (B3 * B10) is computed in cell B18. Cell B10 has the constant for the hours needed to make one basketball. Cell B3 (a changing cell) has the number of basketballs made. (Currently, this cell shows one ball, but that number will change when the Solver works on the problem.)
- TOTAL FOOTBALL HOURS USED: The number of machine hours needed to make all footballs is calculated similarly in cell B19.
- TOTAL MACHINE HOURS USED (BB + FB): The number of hours needed to make both kinds of balls (cell B20) will be a constraint; this value is the sum of the hours just calculated for footballs and basketballs.

Notice that constants in the Excel cell formulas in Figure D-3 are referred to by their cell addresses. Use the cell address of a constant rather than hard-coding a number in the Excel expression. If the number must be changed later, you will have to change it only in the **CONSTANTS** section cell, not in every cell formula in which you used the value.

Notice that you do not calculate the amounts in the changing cells (here, the number of basketballs and footballs to produce). The Solver will compute those numbers. Also, notice that you can use the changing cell addresses in your formulas. When you do that, you assume the Solver has put the optimal values in each changing cell; your expression makes use of those numbers.

Figure D-4 shows the calculated values after Excel evaluates the cell formulas (with 1s in the changing cells).

	A	B
16	**CALCULATIONS**	
17	RATIO OF BASKETBALLS TO FOOTBALLS	1.00
18	TOTAL BASKETBALL HOURS USED	0.50
19	TOTAL FOOTBALL HOURS USED	0.30
20	TOTAL MACHINE HOURS USED (BB + FB)	0.80

FIGURE D-4 CALCULATIONS section cell values

INCOME STATEMENT Section

The target value is calculated in the spreadsheet body in the **INCOME STATEMENT** section. This is the value that the Solver is expected to maximize or minimize. The spreadsheet body can take any form. In this textbook's Solver cases, the spreadsheet body will be an income statement. Figure D-5 shows the skeleton and formulas that you should enter. A discussion of the line item cell formulas follows the figure.

NOTE

Income statement cells were formatted for two decimal places.

	A	B
22	**INCOME STATEMENT**	
23	BASKETBALL REVENUE (SALES)	=B3*B7
24	FOOTBALL REVENUE (SALES)	=B4*B8
25	TOTAL REVENUE	=B23+B24
26	BASKETBALL MATERIALS COST	=B3*B13
27	FOOTBALL MATERIALS COST	=B4*B14
28	COST OF MACHINE LABOR	=B20*B12
29	TOTAL COST OF GOODS SOLD	=SUM(B26:B28)
30	INCOME BEFORE TAXES	=B25-B29
31	INCOME TAX EXPENSE	=IF(B30<=0,0,B30*B9)
32	NET INCOME AFTER TAXES	=B30-B31

FIGURE D-5 INCOME STATEMENT section cell formulas

- REVENUE (cells B23 and B24) equals the number of balls times the respective unit selling price. The number of balls is in the changing cells, and the selling prices are constants.
- MATERIALS COST (cells B26 and B27) follows a similar logic: number of units times unit cost.
- COST OF MACHINE labor is the calculated number of machine hours times the hourly labor rate for machine workers.
- TOTAL COST OF GOODS SOLD is the sum of the cost of materials and the cost of labor.
- This is the logic of income tax expense: If INCOME BEFORE TAXES is less than or equal to zero, the tax is zero; otherwise, the income tax expense equals the tax rate times income before taxes. An =IF() statement is needed in cell B31.

Excel evaluates the formulas. Figure D-6 shows the results (assuming 1s in the changing cells).

	A	B
22	INCOME STATEMENT	
23	BASKETBALL REVENUE (SALES)	14.00
24	FOOTBALL REVENUE (SALES)	11.00
25	TOTAL REVENUE	25.00
26	BASKETBALL MATERIALS COST	2.00
27	FOOTBALL MATERIALS COST	1.25
28	COST OF MACHINE LABOR	8.00
29	TOTAL COST OF GOODS SOLD	11.25
30	INCOME BEFORE TAXES	13.75
31	INCOME TAX EXPENSE	3.85
32	NET INCOME AFTER TAXES	9.90

FIGURE D-6 INCOME STATEMENT section cell values

Constraints

Constraints are rules that the Solver must observe when computing the optimal answer to a problem. Constraints need to refer to calculated values, or to values in the spreadsheet body. Therefore, you must build those calculations into the spreadsheet design so they are available to your constraint expressions. (There is no section on the face of the spreadsheet for constraints. You'll use a separate window to enter constraints.)

Figure D-7 shows the English and Excel expressions for the basketball and football production problem constraints. A discussion of the constraints follows the figure.

Expression in English	Excel Expression
TOTAL MACHINE HOURS >= 39000	B20 >= 39000
TOTAL MACHINE HOURS <= 40000	B20 <= 40000
MIN BASKETBALLS = 30000	B3 >= 30000
MAX BASKETBALLS = 60000	B3 <= 60000
MIN FOOTBALLS = 20000	B4 >= 20000
MAX FOOTBALLS = 40000	B4 <= 40000
RATIO BBs TO FBs-MIN = 1.5	B17 >= 1.5
RATIO BBs TO FBs-MAX = 1.7	B17 <= 1.7
NET INCOME MUST BE POSITIVE	B32 >= 0

FIGURE D-7 Solver Constraint expressions

- As shown in Figure D-7, notice that a cell address in a constraint expression can be a cell address in the **CHANGING CELLS** section, a cell address in the **CONSTANTS** section, a cell address in the **CALCULATIONS** section, or a cell address in the spreadsheet body.
- You'll often need to set minimum and maximum boundaries for variables. For example, the number of basketballs (MIN and MAX) varies between 30,000 and 60,000.
- Often a boundary value is zero because you want the Solver to find a non-negative result. For example, here you want only answers that yield a positive net income. You tell the Solver that the amount in the net income cell must equal or exceed zero so the Solver does not find an answer that produces a loss.
- Machine hours must be shared between the two kinds of balls. The constraints for the shared resource are B20 >= 39000 and B20 <= 40000, where cell B20 shows the total hours used to make both basketballs and footballs. The shared resource constraint seems to be the most difficult kind of constraint for students to master when learning the Solver.
- Marketing wants some product balance between products, so ratios of basketballs to footballs must be in the constraints.

Running the Solver: Mechanics

To set up the Solver, you must tell the Solver:

1. The cell address of the "target" variable that you are trying to maximize (or minimize, as the case may be).
2. The changing cell addresses.
3. The expressions for the constraints.

The Solver will record its answers in the changing cells and on a separate sheet.

Beginning to Set Up the Solver

AT THE KEYBOARD

To start setting up the Solver, first select the Data tab. In the Analysis group, select the Solver. The first thing you see is a Solver Parameters window, as shown in Figure D-8. Use the Solver Parameters window to specify the target cell, the changing cells, and the constraints. (Click the Office Button, then click **Excel Options**, click **Add-Ins**, click the **Go** button for Manage: Excel Add-ins, select the check box for Solver Add-in, then click **OK**. When prompted to Install, click **Yes**.)

FIGURE D-8 Solver Parameters window

Setting the Target Cell

To set a target cell, use the following procedure:

1. Click the **Set Target Cell** box and enter B32 (net income).
2. Accept Max, the default.
3. Enter a zero for no desired net income value (Value of). Do *not* press Enter when you finish. You'll navigate within this window by clicking in the next input box.

Figure D-9 shows the entering of data in the Set Target Cell box.

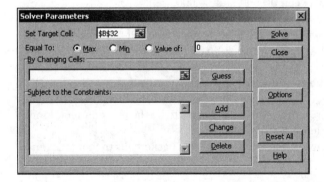

FIGURE D-9 Entering data in the Set Target Cell box

When you enter the cell address, Solver may put in dollar signs, as if for absolute addressing. Ignore them—do not try to delete them.

Setting the Changing Cells

The changing cells are the cells for the balls, which are in the range of cells B3:B4. Click the **By Changing Cells** box and enter B3:B4, as shown in Figure D-10. (Do *not* press Enter.)

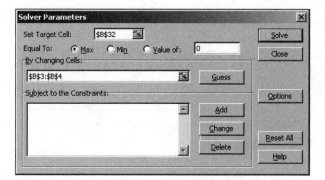

FIGURE D-10   Entering data in the By Changing Cells box

Entering Constraints

You are now ready to enter the constraint formulas one by one. To start, click the **Add** button. As shown in Figure D-11, you'll see the Add Constraint window (shown here with the minimum basketball production constraint entered).

FIGURE D-11   Entering data in the Add Constraint window

You should note the following about entering constraints and about Figure D-11:

- To enter a constraint expression, do four things: (1) Type the variable's cell address in the left Cell Reference input box; (2) select the operator (<=, =, or >=) in the smaller middle box; (3) enter the expression's right-side value, which is either a raw number or the cell address of a value, into the Constraint box; and (4) click **Add** to enter the constraint into the program. If you change your mind about the expression and do not want to enter it, click **Cancel**.
- The minimum basketballs constraint is B3 >= 30000. Enter that constraint now. (Later, Solver may put an equal sign in front of the 30000 and dollar signs in the cell reference.)
- After entering the constraint formula, click the **Add** button. Doing so puts the constraint into the Solver model. It also leaves you in the Add Constraint window, allowing you to enter other constraints. You should enter those now. See Figure D-7 for the logic.
- When you're done entering constraints, click the **Cancel** button. That takes you back to the Solver Parameters window.

Referring again to Figure D-11, you should not put an expression in the Cell Reference box. For example, the constraint for the minimum basketball-to-football ratio is B3/B4 >= 1.5. You should not put =B3/B4 in the Cell Reference box. The ratio is computed in the **CALCULATIONS** section of the spreadsheet (in cell B17). When adding that constraint, enter B17 in the Cell Reference box. (You are allowed to put an expression in the Constraint box, although that technique is not shown here and is not recommended.)

After entering all of the constraints, you'll be back at the Solver Parameters window. You will see that the constraints have been entered into the program. Not all constraints will show, due to the size of the box. The top part of the box's constraints area looks like the portion of the spreadsheet shown in Figure D-12.

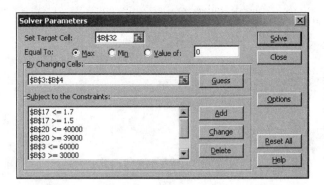

FIGURE D-12 A portion of the constraints entered in the Solver Parameters window

Using the scroll arrow, reveal the rest of the constraints, as shown in Figure D-13.

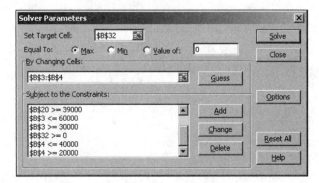

FIGURE D-13 Remainder of constraints entered in the Solver Parameters window

Computing the Solver's Answer

To have the Solver calculate answers, click **Solve** in the upper right corner of the Solver Parameters window. The Solver does its work in the background—you do not see the internal calculations. Then the Solver gives you a Solver Results window, as shown in Figure D-14.

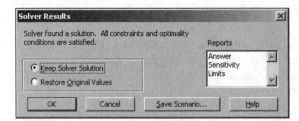

FIGURE D-14 Solver Results window

In the Solver Results window, the Solver tells you that it has found a solution and that the optimality conditions were met. That is a very important message—you should always check for it. It means that an answer was found and the constraints were satisfied.

By contrast, your constraints might be such that the Solver cannot find an answer. For example, suppose you had a constraint that said, in effect "Net income must be at least a billion dollars." That amount cannot be reached, given so few basketballs and footballs and the prices. The Solver would report that no answer is feasible. The Solver may find an answer by ignoring some constraints. Solver would tell you that, too. In either case, there would be something wrong with your model and you would need to rework it.

There are two ways to see your answers. One way is to click **OK**, which lets you see the new changing cell values. A more formal (and complete) way is to click **Answer** in the Reports box and then click **OK**. That puts detailed results into a new sheet in your Excel book. The new sheet is called an Answer Report. All answer reports are numbered sequentially as you run the Solver.

To see the Answer Report, click its tab, as shown in Figure D-15 (here, Answer Report 1).

26	$B$4	NUMBER OF FOOTBALLS	38095.2442 $B$4>=20000 Not Binding
27	$B$4	NUMBER OF FOOTBALLS	38095.2442 $B$4<=40000 Not Binding
28			
29			
30			
31			
32			
33			
34			
35			
36			
37			
38			

Answer Report 1 / Sheet1 / Sheet2 / Sheet3

FIGURE D-15   Answer Report Sheet tab

That takes you to the Answer Report. The top portion of the report is shown in Figure D-16.

	A B	C	D	E	F
1	Microsoft Excel 12.0 Answer Report				
2	Worksheet: [SPORTS1.xlsx]Sheet1				
3					
4					
5					
6	Target Cell (Max)				
7	Cell	Name	Original Value	Final Value	
8	$B$32	NET INCOME AFTER TAXES	9.90	473142.87	
9					
10					
11	Adjustable Cells				
12	Cell	Name	Original Value	Final Value	
13	$B$3	NUMBER OF BASKETBALLS	1	57142.85348	
14	$B$4	NUMBER OF FOOTBALLS	1	38095.2442	

FIGURE D-16   Top portion of the Answer Report

Figure D-17 shows the remainder of the Answer Report.

	A B	C	D	E	F
17	Constraints				
18	Cell	Name	Cell Value	Formula	Status
19	$B$17	RATIO OF BASKETBALLS TO FOOTBALLS	1.50	$B$17<=1.7	Not Binding
20	$B$17	RATIO OF BASKETBALLS TO FOOTBALLS	1.50	$B$17>=1.5	Binding
21	$B$20	TOTAL MACHINE HOURS USED (BB + FB)	40000.00	$B$20<=40000	Binding
22	$B$20	TOTAL MACHINE HOURS USED (BB + FB)	40000.00	$B$20>=39000	Not Binding
23	$B$32	NET INCOME AFTER TAXES	473142.87	$B$32>=0	Not Binding
24	$B$3	NUMBER OF BASKETBALLS	57142.85348	$B$3>=30000	Not Binding
25	$B$3	NUMBER OF BASKETBALLS	57142.85348	$B$3<=60000	Not Binding
26	$B$4	NUMBER OF FOOTBALLS	38095.2442	$B$4>=20000	Not Binding
27	$B$4	NUMBER OF FOOTBALLS	38095.2442	$B$4<=40000	Not Binding

FIGURE D-17   Remainder of the Answer Report

At the beginning of this tutorial, the changing cells had a value of 1 and the income was $9.90 (Original Value). The optimal solution values (Final Value) also are shown: $473,142.87 for net income (the target) and 57,142.85 basketballs and 38,095.24 footballs for the changing (adjustable) cells. (Of course, you cannot make a part of a ball. The Solver can be asked to find only integer solutions; that technique is discussed at the end of this tutorial.)

The report also shows detail for the constraints: the constraint expression and the value that the variable has in the optimal solution. *Binding* means the final answer caused Solver to bump up against the constraint. For example, the maximum number of machine hours was 40,000, which is the value Solver used to find the answer.

Not binding means the reverse. A better word for *binding* might be *constraining*. For example, the 60,000 maximum basketball limit did not constrain the Solver.

The procedures used to change (edit) or delete a constraint are discussed later in this tutorial.

Use the Office Button to print the worksheets (Answer Report and Sheet1). Save the Excel file (Save). Then use Save As to make a new file called SPORTS2.xls to be used in the next section of this tutorial.

EXTENDING THE EXAMPLE

Next, you'll modify the sporting goods spreadsheet. Suppose management wants to know what net income would be if certain constraints were changed. In other words, management wants to play "what if" with certain Base Case constraints. The resulting second case is called the Extension Case. Here are some changes to the original Base Case conditions:

- Assume maximum production constraints will be removed.
- Similarly, the basketball-to-football production ratios (1.5 and 1.7) will be removed.
- There still will be minimum production constraints at some low level. Assume at least 30,000 basketballs and 30,000 footballs will be produced.
- The machine-hours shared resource imposes the same limits as previously.
- A more ambitious profit goal is desired. The ratio of net income after taxes to total revenue should be greater than or equal to .33. This constraint will replace the constraint calling for profits greater than zero.

AT THE KEYBOARD

Begin by putting 1s in the changing cells. You need to compute the ratio of net income after taxes to total revenue. Enter that formula in cell B21. (The formula should have the net income after taxes cell address in the numerator and the total revenue cell address in the denominator.) In the Extension Case, the value of this ratio for the Solver's optimal answer must be at least .33. Click the **Add** button and enter that constraint.

Then in the Solver Parameters window, constraints that are no longer needed are highlighted (click to select) and deleted (click the **Delete** button). Do that for the net income >= 0 constraint, the maximum football and basketball constraints, and the basketball-to-football ratio constraints.

The minimum football constraint must be modified, not deleted. Select that constraint; then click **Change**. That takes you to the Add Constraint window. Edit the constraint so 30,000 is the lower boundary.

When you are finished with the constraints, your Solver Parameters window should look like the one shown in Figure D-18.

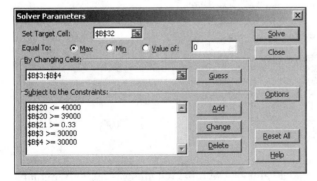

FIGURE D-18 Extension Case Solver Parameters window

You can tell Solver to solve for integer values. Here cells B3 and B4 should be whole numbers. You use the Int constraint to do that. Figure D-19 shows how to enter the Int constraint.

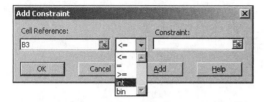

FIGURE D-19 Entering the Int constraint

Make those constraints for the changing cells. Your constraints should now look like the beginning portion of those shown in Figure D-20.

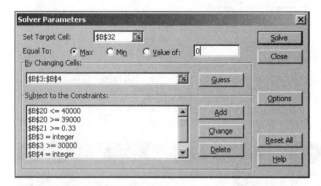

FIGURE D-20 Portion of Extension Case constraints

Scroll to see the remainder of the constraints, as shown in Figure D-21.

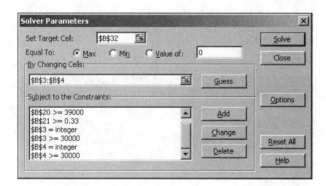

FIGURE D-21 Remainder of Extension Case constraints

The constraints are now only for the minimum production levels, the ratio of net income after taxes to total revenue, machine-hours shared resource constraints, and whole number output. When the Solver is run, the values in the Answer Report look like those shown in Figure D-22.

A	B	C	D	E	F
6	Target Cell (Max)				
7	Cell	Name	Original Value	Final Value	
8	$B$32	NET INCOME AFTER TAXES	9.90	556198.38	
9					
10					
11	Adjustable Cells				
12	Cell	Name	Original Value	Final Value	
13	$B$3	NUMBER OF BASKETBALLS	1	30000	
14	$B$4	NUMBER OF FOOTBALLS	1	83333	
15					
16					
17	Constraints				
18	Cell	Name	Cell Value	Formula	Status
19	$B$20	TOTAL MACHINE HOURS USED (BB + FB)	39999.90	$B$20>=39000	Not Binding
20	$B$21	RATIO OF NET INCOME TO TOTAL REVENUE	0.416109655	$B$21>=0.33	Not Binding
21	$B$20	TOTAL MACHINE HOURS USED (BB + FB)	39999.90	$B$20<=40000	Not Binding
22	$B$3	NUMBER OF BASKETBALLS	30000	$B$3>=30000	Binding
23	$B$4	NUMBER OF FOOTBALLS	83333	$B$4>=30000	Not Binding
24	$B$3	NUMBER OF BASKETBALLS	30000	$B$3=integer	Binding
25	$B$4	NUMBER OF FOOTBALLS	83333	$B$4=integer	Binding

FIGURE D-22 Extension Case Answer Report

The Extension Case answer differs from the Base Case answer. Which production schedule should management use: the one that has maximum production limits or the one that has no such limits? That question is posed to get you to think about the purpose of using a DSS program. Two scenarios, the Base Case and the Extension Case, were modeled in the Solver. The very different answers are shown in Figure D-23.

	Base Case	Extension Case
Basketballs	57,143	30,000
Footballs	38,095	83,333

FIGURE D-23 The Solver's answers for the two cases

Can you use this output alone to decide how many of each kind of ball to produce? No, you cannot. You also must refer to the "Target," which in this case is net income. Figure D-24 shows the answers with net income target data.

	Base Case	Extension Case
Basketballs	57,143	30,000
Footballs	38,095	83,333
Net Income	$473,143	$556,198

FIGURE D-24 The Solver's answers for the two cases—with target data

Viewed this way, the Extension Case production schedule looks better because it gives you a higher target net income.

At this point, you should save the SPORTS2.xls file and then close it (Office Button—Close).

USING THE SOLVER ON A NEW PROBLEM

Here is a short problem that will let you find out what you have learned about the Excel Solver.

Setting Up the Spreadsheet

Assume you run a shirt manufacturing company. You have two products: (1) polo-style T-shirts and (2) dress shirts with button-down collars. You must decide how many T-shirts and how many button-down shirts to make. Assume you'll sell every shirt you make.

AT THE KEYBOARD

Start a new file called SHIRTS.xls. Set up a Solver spreadsheet to handle this problem.

CHANGING CELLS Section

Your changing cells should look like those shown in Figure D-25.

	A	B
1	**SHIRT MANUFACTURING EXAMPLE**	
2	**CHANGING CELLS**	
3	NUMBER OF T-SHIRTS	1
4	NUMBER OF BUTTON-DOWN SHIRTS	1

FIGURE D-25   Shirt manufacturing changing cells

CONSTANTS Section

Your spreadsheet should contain the constants shown in Figure D-26. A discussion of constant cells (and some of your company's operations) follows the figure.

	A	B
6	**CONSTANTS**	
7	TAX RATE	0.28
8	SELLING PRICE: T-SHIRT	8.00
9	SELLING PRICE: BUTTON-DOWN SHIRT	36.00
10	VARIABLE COST TO MAKE: T-SHIRT	2.50
11	VARIABLE COST TO MAKE: BUTTON-DOWN SHIRT	14.00
12	COTTON USAGE (LBS): T-SHIRT	1.50
13	COTTON USAGE (LBS): BUTTON-DOWN SHIRT	2.50
14	TOTAL COTTON AVAILABLE (LBS)	13000000
15	BUTTONS PER T-SHIRT	3.00
16	BUTTONS PER BUTTON-DOWN SHIRT	12.00
17	TOTAL BUTTONS AVAILABLE	110000000

FIGURE D-26   Shirt manufacturing constants

- TAX RATE: The rate is -28 on pretax income, but no taxes are paid on losses.
- SELLING PRICE: You sell polo-style T-shirts for $8 and button-down shirts for $36.
- VARIABLE COST TO MAKE: It costs $2.50 to make a T-shirt and $14 to make a button-down shirt. These variable costs are for machine operator labor, cloth, buttons, etc.
- COTTON USAGE: Each polo T-shirt uses 1.5 pounds of cotton fabric. Each button-down shirt uses 2.5 pounds of cotton fabric.
- TOTAL COTTON AVAILABLE: You have only 13 million pounds of cotton on hand to make all of the T-shirts and button-down shirts.
- BUTTONS: Each polo T-shirt has 3 buttons. By contrast, each button-down shirt has 1 button on each collar tip, 8 buttons down the front, and 1 button on each cuff, for a total of 12 buttons. You have 110 million buttons on hand to be used to make all of your shirts.

CALCULATIONS Section

Your spreadsheet should contain the calculations shown in Figure D-27.

	A	B
19	**CALCULATIONS**	
20	RATIO OF NET INCOME TO TOTAL REVENUE	
21	COTTON USED: T-SHIRTS	
22	COTTON USED: BUTTON-DOWN SHIRTS	
23	COTTON USED: TOTAL	
24	BUTTONS USED: T-SHIRTS	
25	BUTTONS USED: BUTTON-DOWN SHIRTS	
26	BUTTONS USED: TOTAL	
27	RATIO OF BUTTON-DOWNS TO T-SHIRTS	

FIGURE D-27 Shirt manufacturing calculations

Calculations (and related business constraints) are discussed next.

- RATIO OF NET INCOME TO TOTAL REVENUE: The minimum return on sales (ratio of net income after taxes divided by total revenue) is .20.
- COTTON USED/BUTTONS USED: You have a limited amount of cotton and buttons. The usage of each resource must be calculated, then used in constraints.
- RATIO OF BUTTON-DOWNS TO T-SHIRTS: You think you must make at least 2 million T-shirts and at least 2 million button-down shirts. You want to be known as a balanced shirtmaker, so you think the ratio of button-downs to T-shirts should be no greater than 4:1. (Thus, if 9 million button-down shirts and 2 million T-shirts were produced, the ratio would be too high.)

INCOME STATEMENT Section

Your spreadsheet should have the income statement skeleton shown in Figure D-28.

	A	B
29	**INCOME STATEMENT**	
30	T-SHIRT REVENUE	
31	BUTTON-DOWN SHIRT REVENUE	
32	TOTAL REVENUE	
33	VARIABLE COSTS: T-SHIRTS	
34	VARIABLE COSTS: BUTTON-DOWNS	
35	TOTAL COSTS	
36	INCOME BEFORE TAXES	
37	INCOME TAX EXPENSE	
38	NET INCOME AFTER TAXES	

FIGURE D-28 Shirt manufacturing income statement line items

The Solver's target is net income, which must be maximized.

Use the table shown in Figure D-29 to write out your constraints before entering them into the Solver.

Expression in English	Fill in the Excel Expression
Net income to revenue	_____ >= _____
Ratio of BDs to Ts	_____ <= _____
Min T-shirts	_____ >= _____
Min button-downs	_____ >= _____
Usage of buttons	_____ <= _____
Usage of cotton	_____ <= _____

FIGURE D-29 Logic of shirt manufacturing constraints

When you are finished with the program, print the sheets. Then use the Office Button to Save the file, Close it, and choose Exit to leave Excel.

TROUBLESHOOTING THE SOLVER

Use this section to overcome problems with the Solver and to review of some Windows file-handling procedures.

Rerunning a Solver Model

Assume you have changed your spreadsheet in some way and want to rerun the Solver to get a new set of answers. (For example, you may have changed a constraint or a formula in your spreadsheet.) Before you click Solve to rerun the Solver, put the number 1 in the changing cells.

Creating Overconstrained Models

It is possible to set up a model that has no logical solution. For example, in the second version of the sporting goods problem, suppose you had specified that at least 1 million basketballs were needed. When you clicked Solve, the Solver would have tried to compute an answer, but then would have admitted defeat by telling you that no feasible solution was possible, as shown in Figure D-30.

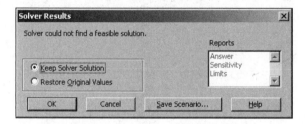

FIGURE D-30 Solver Results message: Solution not feasible

In the Reports window, the choices (Answer, Sensitivity, and Limits) would be in gray, indicating that they are not available as options. Such a model is sometimes referred to as overconstrained.

Setting a Constraint to a Single Amount

You may want an amount to be a specific number, as opposed to a number in a range. For example, if the number of basketballs needed to be exactly 30,000, then the "equals" operator would be selected, as shown in Figure D-31.

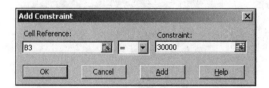

FIGURE D-31 Constraining a value to equal a specific amount

Setting a Changing Cell to an Integer

You may want to force changing cell values to be integers. The way to do that is to select the Int operator in the Add Constraint window. That was described in a prior section.

Forcing the Solver to find only integer solutions slows down the Solver. In some cases, the change in speed can be noticeable to the user. Doing this also can prevent the Solver from seeing a feasible solution when in fact one can be found if the Solver is allowed to find noninteger answers. For those reasons, it's usually best not to impose the integer constraint unless the logic of the problem demands it.

Deleting Extra Answer Sheets

Suppose you've run different scenarios, each time asking for an Answer Report. As a result, you have a number of Answer Report sheets in your Excel file, but you don't want to keep them all. How do you get rid of an Answer Report sheet? Follow this procedure: First, select the Home tab. In the Cells group, click the **Delete** drop-down arrow and select Delete Sheet. You will *not* be asked if you really mean to delete. Therefore, make sure you want to delete the sheet before you act.

Restarting the Solver with All New Constraints

Suppose you wanted to start over with a new set of constraints. In the Solver Parameters window, click **Reset All**. You will be asked if you really mean to do that, as shown in Figure D-32.

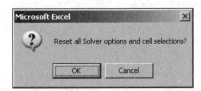

FIGURE D-32   Reset options warning query

If you do, select OK. That gives you a clean slate, with all entries deleted, as shown in Figure D-33.

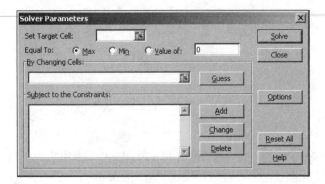

FIGURE D-33   Reset Solver Parameters window

As you can see, the target cell, changing cells, and constraints have been reset. From this point, you can specify a new model.

NOTE

If you select Reset All, you really are starting over. If you merely want to add, delete, or edit a constraint, do not use Reset All. Use the Add, Delete, or Change buttons as appropriate.

Using the Solver Options Window

The Solver has a number of internal settings that govern its search for an optimal answer. If you click the Options button in the Solver Parameters window, you will see the defaults for those settings, as shown in Figure D-34.

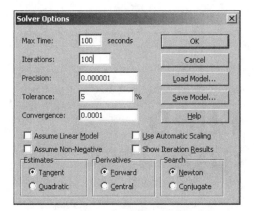

FIGURE D-34 Solver Options window, with default settings for Solver Parameters

In general, Solver Options govern how long the Solver works on a problem and/or how precise it must be in satisfying constraints. You should not check Assume Linear Model if changing cells are multiplied or divided (as they are in this book's cases) or if some of the spreadsheet's formulas use exponents.

You should not have to change these default settings for the cases in this book. If you think that your Solver work is correct but Solver cannot find a feasible solution, you should check to see that Solver Options are set as shown in Figure D-34.

Printing Cell Formulas in Excel

To show the Cell Formulas on the screen, press Ctrl and the left quote (') keys at the same time: Ctrl-'. (The left quote is usually on the same key as the tilde (~).) Pressing Ctrl-' automatically widens cells so the formulas can be read. You can change cell widths by clicking and dragging at the column indicator (A, B, C, etc.) boundaries.

To print the formulas, use the Office Button and select Print. Print the sheet as you would normally. To restore the screen to its typical appearance (showing values, not formulas), press Ctrl-' again. (It's like a toggle switch.) If you did not change any column widths while in the cell formula view, the widths will be as they were.

Reviewing the Printing, Saving, and Exiting Procedures

Print the Solver spreadsheets in the normal way. Activate the sheet; then select the Office Button and Print. You can print an Answer Report sheet the same way.

To save a file, use the Office Button and Save (or Save As). Make sure you select the proper drive, for example, drive A: if you intend your file to be on a disk. When exiting from Excel, always start with the Office Button. Then Close (with the disk in drive A: if your file is on the disk) and select Exit. Only then should you take the disk out of drive A:.

> **NOTE**
>
> If you use File—Exit (without selecting Close first), you risk losing your work.

Sometimes the Solver will come up with strange results. For instance, your results might differ from the target answers that your instructor provides for a case. Thinking that you've done something wrong, you ask to compare your cell formulas and constraint expressions with those your instructor created. Lo and behold, you see no differences! Surprisingly and for no apparent reason, Solver occasionally produces slightly different outputs from inputs that are seemingly the same. The reason may be because for your application, the order of the constraints (or even the order in which they are entered) matters. In any case, if you are close to the target answers but cannot see any errors, it's best to see your instructor for guidance, rather than to spin your wheels.

Here is another example. Assume you ask for Integer changing cell outputs. The Solver may tell you that the correct output is 8.0000001, or 7.9999999. In such situations, the Solver is apparently not sure about its own rounding. In that case, you should merely humor the Solver and (continuing the example) take the result as the integer 8, which is what the Solver was trying to say in the first place.

THE COLLEGE MUTUAL FUND INVESTMENT MIX DECISION

Decision Support Using Excel

PREVIEW

A college's MBA program runs a mutual fund so students can learn how to invest money. In this case, you will use the Excel Solver to give the students advice on the best mix of securities to buy.

PREPARATION

Before attempting this case, you should:

- Review spreadsheet concepts discussed in class and/or in your textbook.
- Complete any exercises that your instructor assigns.
- Complete any part of Tutorial D that your instructor assigns or refer to it as necessary.
- Review file-saving procedures for Windows programs, which are discussed in Tutorial C.
- Refer to Tutorial F as necessary.

BACKGROUND

Old Oak College is a prestigious private college on the outskirts of your state's capital. Old Oak's MBA program is highly respected in part because it strives to teach students about investing money in the real world. In their final year, six finance majors are chosen by their professors to run their own small mutual fund. The students are given control over $1 million of Old Oak's massive endowment. They invest the money in different kinds of stocks. At the end of the school year, the money is returned to the Old Oak endowment—with the hope that the students have earned a profit, thereby increasing the endowment's value. At the start of the next school year, six new MBA students get a million dollars to invest the same way.

By definition, a mutual fund is a company that invests in the stock of other companies. If the market value of the other companies' stock goes up, the value of the mutual fund goes up. If the value of the other companies' stock goes down, the value of the mutual fund goes down. Also, if the other companies pay a dividend, the cash goes to the mutual fund, thus increasing the value of the mutual fund's shares. Mutual funds also can buy bonds or other securities that pay interest. When the interest is paid, the fund's value increases by the cash received.

Old Oak students get investment advice from the huge mutual fund company Felicity, whose founder is an Old Oak alum. Felicity does not give away this advice for free, but the fees it charges Old Oak are low.

This year the MBA students have decided to invest in only certain kinds of securities. Those securities are discussed next, in order of increasing riskiness.

- *Fixed income securities:* These are bonds and other kinds of interest-bearing securities. Another word for this kind of security is *bonds*. The securities are issued by the U.S. government or by stable corporations. These securities pay interest at a fixed rate. The government and the companies are likely to meet their obligations, so the risk of losing money is very low.
- *Value stocks*: These securities are shares of large successful companies with proven track records, for example, shares of companies that are doing well in the chemical or steel industries. There is less risk in this kind of stock than in growth stocks.
- *Growth stocks*: These are shares of companies that are growing rapidly. Examples are shares of drug companies and telecommunication companies. The value of these shares will go up a great deal if

the companies' new business ventures succeed. However, the ventures might not succeed. In that case, the value of the shares go down; so there is some risk in buying these kinds of stock.

- *Gold stocks*: These are shares of companies that are involved in gold mining or gold processing. The value of gold fluctuates greatly in the international commodities markets, so the value of shares of companies in the gold business can go up or down greatly in a short period of time. The shares of these stocks are riskier than any of the other three investments.

Notice that the four kinds of investments differ greatly in their rewards and in their risks. If the value of the shares of one kind goes up, the value of the shares of at least one of the other kinds will probably go down. Buying different kinds of investments is called *diversification*—it is the financial equivalent of "not putting your eggs in one basket." Diversification is a key strategy in building a good portfolio. The likely rate of return and the risk level of each kind of investment are discussed next.

Rates of Return and Risk

Each kind of security has an expected rate of return and an expected level of risk. Assume a rate of return is one of the following:

- The amount of cash received by the investor (in dividends or interest) divided by the cost of the investment
- The increase in the security's market price divided by the cost of the investment

Generally speaking, investors seeking higher rates of return must assume higher levels of risk, including the risk of losing principal. A loss of principal occurs when the value of the security goes down after the security has been purchased. For example, suppose an investor buys 1,000 shares of AOL stock at $80 per share and then the price falls to $10 per share. The loss of principal is 1,000 times $70, or a loss of $70,000.

Old Oak's finance students have estimated the expected annual rate of return for the four kinds of securities under consideration. The rates are shown in Figure 8-1.

Type of Security	Expected Annual Rate of Return
Fixed income securities	.03
Value stocks	.04
Growth stocks	.05
Gold stocks	.08

FIGURE 8-1 Expected annual rates of return

For example, suppose $1 million was invested in bonds. By the students' estimates, 3 percent of that amount—$30,000—will be received on that investment after one year. That $30,000 will be "revenue" in the Old Oak fund's income statement. Another example: Suppose $1 million was invested in growth stocks. Figure 8-1 shows that the prices of those stocks are expected to go up an average of 5 percent per year, which would produce $50,000 of revenue in the Old Oak fund's income statement. Notice that the rates of return cited here are annual rates.

The students also have quantified the expected levels of risk for the four kinds of investments. Risk is denoted by an expected risk indicator, which is an integer. The higher the indicator number, the higher the expected risk. The risk indicators are shown in Figure 8-2.

Type of Security	Expected Risk Indicator
Fixed income securities	1
Value stocks	3
Growth stocks	4
Gold stocks	8

FIGURE 8-2 Expected risk indicators

As shown in Figure 8-2, fixed income securities have the lowest risk, because the U.S. government and major corporations can be expected to repay their debts. Value stock companies carry some risk, but not as much as growth stock companies. Gold stock companies are the riskiest—there is some chance that gold prospecting companies will actually become insolvent, resulting in a complete loss of the investment in the company's stock.

Fund Expenses

Running the Old Oak fund will involve some expenses. Those expenses are discussed next.

The MBA program's faculty is concerned that the students will overcommit their time to this interesting project and that their other school work will suffer. Thus, each student is required to spend no more than 16 hours a week helping to manage the fund. (Note that the school year is 40 weeks, not 52 weeks.) Faculty advisers came up with the 16-hour rule many years ago, and it is a long-standing aspect of the mutual fund project.

In managing the fund, students spend their time buying (and selling if necessary) each type of security. For example, the students might choose to buy $250,000 of each kind of security, for a total of $1,000,000. During the school year, Old Oak students also analyze the business activities of the companies, using the techniques they learned in their accounting and finance classes. As they do this, they work with Felicity researchers and traders.

The table in Figure 8-3 shows the hours needed (per $1,000 invested) to keep track of each type of investment.

Type of Security	Hours per $1,000 Invested
Fixed income securities	3
Value stocks	4
Growth stocks	5
Gold stocks	6

FIGURE 8-3 Hours needed to keep track of investments

For an example on how to use this table, suppose the students purchase $200,000 of fixed income securities. According to Figure 8-3, the students are expected to spend three hours per $1,000 invested in analyzing and keeping track of an investment in fixed income securities. That means the students will collectively, over the course of the school year, spend a total of 600 hours (3 * ($200,000/1,000)) on this single investment.

For their work, students will be paid the princely sum of $2 an hour by the Fund. That amount will come out of the revenue received by the Fund.

Although Felicity does not want to advise the Old Oak students for free, the company is willing to work for less than the normal industry rates. Felicity will charge a "management fee" of one-half of 1 percent of the value of the total dollars invested. Example: For the $1,000,000 that is invested, Felicity will charge $5,000 ($1,000,000 * .005).

Likewise, Felicity does not want to give away for free the time its employees spend on helping the students. Therefore, a low "consulting expense" per $1,000 invested also will be charged. The rates for that consulting expense are shown in Figure 8-4.

Type of Security	Dollars per $1,000 Invested
Fixed income securities	10
Value stocks	15
Growth stocks	20
Gold stocks	25

FIGURE 8-4 Consulting charge for each type of investment

Return to the example of a $200,000 investment in fixed income securities. The consulting fee for the year will be as follows: $10 * ($200,000/$1,000) = $2,000.

The challenge for the six Old Oak students is to decide how many dollars to invest in each kind of security during the year. They think they should invest at least $150,000 in each. To be conservative, they have adopted this rule: fixed income securities plus value stocks must be at least 60 percent of the total dollars invested. Also, the students have pledged to observe the 16-hour rule, which means they will have to budget their time as carefully as their investments. That will be difficult as the students want to do all they can to maximize their fund's net income after taxes.

The students can invest up to $1 million. The students must invest at least $980,000, leaving $20,000 in cash on the side. The reason is that stock share prices may not evenly divide into the amounts allocated, so investing exactly $1 million would probably not be possible. (Note that cash earns no return and has no risk, so it need not be included as a security in the analysis.)

A couple of the six MBA students think that the 16-hour rule is too restrictive. That minority also thinks that investment safety (the risk(s) associated with investments) should be assessed in more than one way.

Because the students heard that you know how to use the Excel Solver to help with this sort of problem, they have called you in. You must model two different situations: a Base Case and an alternative Extension Case. After you have finished the Base Case, you will modify its spreadsheet to create the Extension Case. In the Extension Case, you will modify the 16-hour rule and use an alternative investment safety rule. Comparing the results of the two cases will let you play out some "what-if" scenarios with the decision. Old Oak students will use your results to decide how to structure their portfolio.

ASSIGNMENT 1: CREATING A SPREADSHEET FOR DECISION SUPPORT

In this assignment, you will produce a spreadsheet that models the business decision. In Assignment 1A, you will make a Solver spreadsheet to model the Base Case. In Assignment 1B, you will make a Solver spreadsheet to model the Extension Case. In Assignment 2, you will use the spreadsheet models to develop the information needed to recommend the best investment mix. Then you will document your recommendation in a memo to the Old Oak students. In Assignment 3, you will give your recommendation in an oral presentation.

First, you must create the spreadsheet models of the decision. Your spreadsheets should have the following sections:

- **CHANGING CELLS**
- **CONSTANTS**
- **CALCULATIONS**
- **INCOME STATEMENT**

Your spreadsheets also will include the decision constraints, which will be discussed later in this case.

A Base Case spreadsheet skeleton is available to you, so you do not need to type it in. To access the spreadsheet skeleton, go to your Data files, select Case 8, and then select OLDOAK1.xls.

Assignment 1A: Creating the Spreadsheet—Base Case

You will model the investment problem described. Your model, when run, will tell the students how to invest in each kind of security during the year.

A discussion of each spreadsheet section follows. For each section, you will learn (1) how the section should be set up and (2) the logic of the sections' cell formulas.

CHANGING CELLS Section

Your spreadsheet should have the changing cells shown in Figure 8-5.

In this section, you are asking the Solver model to compute the number of dollars that should be invested in each kind of security during the school year. Start with a "$1" in each cell. The Solver will change each 1 as it computes the answer. *Note:* It is conceivable that the students' investments could have values that include fractions of a dollar; for example, $250,100.**50**.

	A	B	C
1	OLD OAK COLLEGE MUTUAL FUND		
2			BASE
3			CASE
4	CHANGING CELLS		
5	FIXED INCOME SECURITIES		$1.00
6	VALUE STOCKS		$1.00
7	GROWTH STOCKS		$1.00
8	GOLD STOCKS		$1.00

FIGURE 8-5 CHANGING CELLS section

CONSTANTS Section

Your spreadsheet should have the constants shown in Figure 8-6. An explanation of the line items follows the figure.

	A	B	C
10	CONSTANTS		
11	TAX RATE		0.3
12	NUMBER OF STUDENTS ON TEAM		6
13	RATE OF RETURN:		--
14	FIXED INCOME SECURITIES		0.03
15	VALUE STOCKS		0.04
16	GROWTH STOCKS		0.05
17	GOLD STOCKS		0.08
18	STUDENT ANALYSIS HOURS PER $1000 PER YEAR		--
19	FIXED INCOME SECURITIES		3
20	VALUE STOCKS		4
21	GROWTH STOCKS		5
22	GOLD STOCKS		6
23	CONSULTING EXPENSE PER $1000		--
24	FIXED INCOME SECURITIES		10
25	VALUE STOCKS		15
26	GROWTH STOCKS		20
27	GOLD STOCKS		25
28	MANAGEMENT FEE %		0.005
29	HOURS AVAILABLE PER WEEK PER STUDENT		16

FIGURE 8-6 CONSTANTS section

- TAX RATE: Income tax expense is 30 percent of the net income before taxes.
- NUMBER OF STUDENTS ON TEAM: Six students are on the Old Oak mutual fund team.
- RATE OF RETURN: As Figure 8-6 shows, this rate differs for each kind of investment.
- STUDENT ANALYSIS HOURS PER $1,000 PER YEAR: As shown in Figure 8-6, the number of hours students are expected to spend analyzing their investments differs for each kind of investment.
- CONSULTING EXPENSE PER $1,000: As shown in Figure 8-6, Felicity charges a different fee for each kind of investment.
- MANAGEMENT FEE %: Felicity charges an overall management fee of one-half of 1 percent of the total dollars invested.
- HOURS AVAILABLE PER WEEK PER STUDENT: Each of the six students can work on the fund for 16 hours per week during the school year. Assume students will not work on the fund during the other weeks in the calendar year.

CALCULATIONS Section

Your spreadsheet should calculate the amounts shown in Figure 8-7. They will be used in the **INCOME STATEMENT** section and/or the CONSTRAINTS. Calculated values are based on changing cell values, on constants, and/or on other calculated values. An explanation of the line items follows the figure.

- REVENUE: Calculate the revenue for each kind of investment. Revenue is a function of the amount invested during the year and the annual rate of return.
- STUDENT ANALYSIS HOURS IN SCHOOL YEAR: For each kind of investment, compute the analysis hours required. This is a function of the amount invested and the hours required per $1,000 invested.

	A	B	C
31	**CALCULATIONS**		
32	REVENUE:		--
33	FIXED INCOME SECURITIES		
34	VALUE STOCKS		
35	GROWTH STOCKS		
36	GOLD STOCKS		
37	STUDENT ANALYSIS HOURS IN SCHOOL YEAR:		--
38	FIXED INCOME SECURITIES		
39	VALUE STOCKS		
40	GROWTH STOCKS		
41	GOLD STOCKS		
42	STUDENT ANALYSIS EXPENSE:		--
43	FIXED INCOME SECURITIES		
44	VALUE STOCKS		
45	GROWTH STOCKS		
46	GOLD STOCKS		
47	CONSULTING EXPENSE:		--
48	FIXED INCOME SECURITIES		
49	VALUE STOCKS		
50	GROWTH STOCKS		
51	GOLD STOCKS		
52	TOTAL INVESTED		
53	MANAGEMENT FEE		
54	TOTAL INVESTED IN BONDS + VALUE STOCKS		
55	PERCENT INVESTED IN BONDS + VALUE STOCKS		
56	TOTAL STUDENT ANALYSIS HOURS USED		
57	STUDENT ANALYSIS HOURS AVAILABLE		

FIGURE 8-7 CALCULATIONS section

- STUDENT ANALYSIS EXPENSE: This is a function of the student analysis hours during the year and the rate per hour, which is $2.
- CONSULTING EXPENSE: For each kind of investment, compute the consulting expense that Felicity will charge. This is a function of the amount invested and the consulting expense charged per $1,000 invested.
- TOTAL INVESTED: This is the total of the amounts that were computed by the Solver and shown in the **CHANGING CELLS** section.
- MANAGEMENT FEE: This is a function of the total dollars invested and Felicity's management fee, which is one-half of 1 percent.
- TOTAL INVESTED IN BONDS + VALUE STOCKS: This is the sum of the amounts invested in fixed income securities and value stocks.
- PERCENT INVESTED IN BONDS + VALUE STOCKS: This is the ratio of the total invested in fixed income securities and value stocks to total dollars invested.
- TOTAL STUDENT ANALYSIS HOURS USED: This is the total of student analysis hours used during the school year.
- STUDENT ANALYSIS HOURS AVAILABLE: This is a function of the number of students, the length of the school year in weeks, and the hours allowed to each student in a week.

INCOME STATEMENT Section

Compute the net income of the Old Oak fund, as shown in Figure 8-8. An explanation of the line items follows the figure.

	A	B	C
60	**INCOME STATEMENT**		
61	REVENUE		
62	MANAGEMENT FEE		
63	STUDENT ANALYSIS EXPENSE		
64	CONSULTING EXPENSE		
65	TOTAL EXPENSES		
66	INCOME BEFORE INCOME TAXES		
67	INCOME TAX EXPENSE		
68	NET INCOME AFTER TAXES		

FIGURE 8-8 INCOME STATEMENT section

- REVENUE: This is the sum of revenues calculated for each kind of investment.
- MANAGEMENT FEE: This is a calculation of Felicity's management, which can be echoed here.
- STUDENT ANALYSIS EXPENSE: This is the sum of analysis expenses calculated for each kind of investment.
- CONSULTING EXPENSE: This is the sum of consulting expenses calculated for each kind of investment.
- TOTAL EXPENSES: This is the sum of the management fee, student analysis expense, and consulting expense.
- INCOME BEFORE INCOME TAXES: This is the difference between Revenue and Total Expenses.
- INCOME TAX EXPENSE: Taxes are zero when Income Before Income Taxes is zero or negative. Otherwise, apply the tax rate (a constant) to the Income Before Income Taxes to compute the income tax.
- NET INCOME AFTER TAXES: This is the difference between Income Before Income Taxes and the Income Tax Expense.

Constraints and Running the Solver

In this part of the assignment, you will determine the decision constraints for the spreadsheet. Enter the Base Case decision constraints using the Solver. Run the Solver; when the Solver says that a solution has been found that satisfies the constraints, ask for the Answer Report.

When you are done, print the entire workbook, including the Answer Report sheet. Save the workbook one last time (Office Button—Save). Then to prepare for the Extension Case, use Office Button—Save As to make a new spreadsheet file. (OLDOAK2.xls would be a good name.)

Assignment 1B: Creating the Spreadsheet—Extension Case

A few people on the student team who call themselves the "hard-working conservatives" want to (1) rethink the 16-hour rule and (2) include an investment policy with added safety.

The hard-working conservatives think that the 16-hours-per-week rule is too restrictive. They believe that if students can spare the time (and net income after taxes would be greatly increased), then why shouldn't they be allowed to put in extra analysis hours? According to them, the 16-hours-a-week rule simply should not apply.

Those students also do not think that enough safety is provided by the rule stipulating that 60 percent of the investment should comprise fixed income securities and value stocks. They want to observe that rule, but also implement a rule based on each security's risk factors. They propose that a weighted average risk level be computed for the entire fund. That weighted average should not exceed 4—a ceiling that students believe should, when used in conjunction with the 60 percent rule, provide sufficient safety.

The weighted average risk level would be computed as follows:

- Weight investment risk factors by the amount invested in each kind of investment.
- Add the weighted amounts.
- Divide that total by the total amount invested.

Here is an example. Assume that $250,000 is invested in each kind of security. As you may recall from Figure 8-2, the risk factors are fixed income securities:1; value stocks: 3; growth stocks: 4; and gold stocks: 8. The weighted average risk level would equal ($250,000 * 1 + $250,000 * 3 + $250,000 * 4 + $250,000 * 8) / $1,000,000 = 4. Note that this allocation would be acceptable only under the new rule.

You should include the investments' risk factors in the **CONSTANTS** section. The easiest way to do that is to put the risk factors to the right of the Rates of Return, as shown in Figure 8-9.

D	E
RISK FACTOR	
FIXED INCOME SECURITIES	1
VALUE STOCKS	3
GROWTH STOCKS	4
GOLD STOCKS	8

FIGURE 8-9 Risk factors added to CONSTANTS section

Modify the Extension Case spreadsheet and related constraints to reflect the hard-working conservatives' way of thinking. Run the Solver. Ask for an Answer Report when the Solver says that a solution has been found that satisfies the constraints. When you are done, print the entire workbook, including the Solver Answer Report. Save the file when you are done, close the file, and exit Excel.

ASSIGNMENT 2: USING THE SPREADSHEET FOR DECISION SUPPORT

You have built the Base Case and Extension Case models because you want to know the profitability of each investment approach. Now you will complete your work by (1) using the worksheets and Answer Reports to gather data and (2) documenting your analysis and recommendation in a memo.

Assignment 2A: Using the Spreadsheets to Gather Data

Since you printed the spreadsheet and Answer Report for each case, you can see the results of each investment approach. The Old Oak's students are interested in knowing how net income after taxes differs with each model. They also are interested in knowing how many hours of student analysis are required in each case. The Extension Case is likely to call for the students to put in more analysis hours—but its net income is probably better and the added time is probably not too burdensome.

You should summarize the key data in a table that you will include with your memo. The form of that table is shown in Figure 8-10.

	Base Case	Extension Case
Fixed Income Securities		
Value Stocks		
Growth Stocks		
Gold Stocks		
Net Income After Taxes		
Total Student Analysis Hours Used		
Average Student Analysis Hours Used per Week		

FIGURE 8-10   Format of table to insert into memo

The amount of money invested in the different kinds of securities can be found in the Answer Reports and in the worksheet changing cells. Net income after taxes can be found in the Answer Reports and in the worksheets. Total student analysis hours used can be found in the Base Case Answer Report and in the Extension Case worksheet. You should be able to compute average student analysis hours used per week, which is based on total student hours worked, the number of students, and the number of weeks in the school year, by using a calculator.

Assignment 2B: Documenting Your Recommendation in a Memo

Use Word to write a brief memo to the Old Oak students about the results of your analysis. Observe the following requirements:

- Include a proper heading (TO/FROM/DATE/SUBJECT). You may use a Word memo template (Click **Office Button**, click **New**, click the **Memos** template in the **Templates** section, choose **Contemporary Memo**, then click **OK**. The first time you do this you may need to click **Download** to install the template).
- Briefly state the problem and the decision to be made, but do not provide much background—you can assume the students are aware of the problem. Briefly state your analytical method and provide the results of the analysis.
- Give the students your recommendation, which should be based on the highest net income after taxes and on the hours worked. Is it in fact true that extra work leads to extra net income? If so, does the extra work seem to be worthwhile? That is, is the resultant increase in net income significant?

- Support the recommendation graphically by including a summary table in your memo, as shown in Figure 8-10. Create the table by following the procedure described next.

Enter the table into the memo using the following procedure:

1. Click the cursor where you want the table to appear in the document.
2. In the **Insert** group, select the **Table** drop-down menu.
3. Select **Insert Table**.
4. Choose the number of rows and columns.
5. Click **OK**.

ASSIGNMENT 3: GIVING AN ORAL PRESENTATION

Your instructor may request that you present your analysis and recommendations in an oral presentation. If so, assume the students are very impressed by your analysis and they want you to give a presentation to some of their favorite professors and the dean in charge of the MBA program. Prepare to explain your analysis and recommendation to the group in ten minutes or less. Use visual aids or handouts that you think are appropriate. (Tutorial F has information on how to prepare and give an oral presentation.)

DELIVERABLES

Assemble the following deliverables for your instructor:

1. Printout of your memo
2. Spreadsheet printouts
3. Disk that contains your memo and spreadsheet files. Do not provide a CD, as it would be read-only.

Staple the printouts together, placing the memo on top.

SCHEDULING GRILL MANUFACTURING AT WOMBAT GRILLS

Decision Support Using Excel

PREVIEW

Wombat Grill Company needs an Excel Solver program that helps managers schedule production of the company's popular charcoal and gas grills.

PREPARATION

Before attempting this case, you should:

- Review spreadsheet concepts discussed in class and/or in your textbook.
- Complete any exercises that your instructor assigns.
- Complete any part of Tutorial D that your instructor assigns or refer to it as necessary.
- Review file-saving procedures for Windows programs, which are discussed in Tutorial C.
- Refer to Tutorial F as necessary.

BACKGROUND

Winifred Pugh has always enjoyed barbecuing. A decade ago she quit her bartending job to pursue her dream: to make a better outdoor grill for the American people. She founded a company to do just that and named it Wombat Grills. Her banker asked her why she wanted to name the company after the odd-looking wombat. Winnie replied, "Because wombats are good at burrowing into the ground. That is what I want to do: burrow into the barbecue grill market." Winnie's response was good enough for the banker, who agreed to finance the start-up.

Winnie was convinced that grill manufacturers were making things too difficult for the consumer. There were too many grill models and options for the average person to comprehend, and that was depressing sales. The key, she thought, was to offer a few reasonably priced models that worked well. She believed people would line up to buy grills that they could operate easily and that they could afford. And she predicted that her new grills would penetrate—just like a wombat—into existing markets and even expand them. Fortunately, Winnie was right. Her company is doing just fine.

A charcoal grill has a container filled with charcoal briquettes. The briquettes burn at a very hot temperature without much smoke. A liquid propane (LP) gas grill has one or more units that burn propane gas. LP grills cook faster and are easier to use than charcoal grills, but some people think that charcoal grilling produces a better taste. Wombat Grills makes only three models: one charcoal grill and two LP gas grills. The low number of models keeps manufacturing and marketing simple, which reduces costs.

Winnie pioneered the use of interchangeable parts in grill manufacturing. All grills must have a grill body that acts as a frame for the rest of the components. Winnie designed a frame in the form of a rolling cart. She calls this component the cart frame. Each of her three models uses the same cart frame. Using the same frame reduces manufacturing and procurement costs in the same way that an automobile manufacturer reduces costs by putting several different car models on the same frame.

In the same vein, each of Wombat Grill's three models uses the same container, which is a bowl-like component that, depending on which grill it is fitted to, holds the charcoal or houses the LP burner. Each of the three models also uses the same hood, which is the hinged cover that keeps the heat in the container during the cooking process.

In manufacturing, a bill of materials is a list of a product's components. In a sense, the bill of materials is the product's recipe. Figure 9-1 shows the bill of materials for the three Wombat models.

FIGURE 9-1 Bill of materials for three Wombat Grills

Component	Charcoal	Basic LP	Deluxe LP
Cart frame	1	1	1
Hood	1	1	1
LP burner	0	1	2
Auxiliary item pack	0	1	2
Valve regulator	0	1	2
Container	1	1	1
Cardboard box	1	1	1

A charcoal grill has a cart frame, a hood, and a container; and it is packaged in a cardboard box. A basic LP grill has those three components as well as one LP burner, an auxiliary item pack, and a burner valve regulator. The regulator is the knob that controls how much gas is burned and, thus, the amount of heat generated. An auxiliary item pack is a plastic pouch containing nuts and bolts, gas hoses, and a starter unit, which is a loud-clicking igniter that fires up the burner. The deluxe LP grill is a basic LP grill that has an extra burner along with its corresponding auxiliary item pack and value regulator.

Not much direct labor is involved in getting grills ready for the customer. Each kind of grill is a different color, but a Wombat employee can quickly apply the necessary spray painting. In addition to the painting, a Wombat employee applies different detailing and decals for each kind of grill. Then the worker bundles the components and a sheet of assembly directions into a cardboard box; those boxes are stored in the warehouse prior to shipment. (Before using the grill for the first time, customers must put the grill together—in other words, some assembly is required.) Time-and-motion studies show that, on average, Wombat workers spend 4 minutes on a charcoal grill, 9 minutes on a basic LP grill, and 11 minutes on a deluxe LP grill.

Wombat concentrates its grill assembly into certain periods during the year. The company is now entering a grill assembly season, which lasts 16 weeks. For this upcoming assembly period, Winnie has hired 50 workers. Those workers are expected to work eight hours a day; they are paid $10 an hour.

Some administrative costs are involved in manufacturing the grills. Wombat accountants say that for each grill produced, $1 is expended for insurance, office operations, and warehouse operations.

The number of components on hand and the unit cost of each component at the start of the manufacturing period are shown in Figure 9-2.

FIGURE 9-2 Number of components on hand and item cost

Component	Number on Hand	Item Cost
Cart frame	50,000	$50
Hood	50,000	$20
LP burner	40,000	$75
Auxiliary item pack	40,000	$35
Valve regulator	40,000	$30
Container	50,000	$50
Cardboard box	50,000	$2

In the upcoming selling season, the marketing department (Winnie and her son Beux) plans to sell the grills to distributors at the following prices: charcoal grill—$300; basic LP grill—$600; deluxe LP grill—$800.

Winnie is sure she can sell all of the grills that are produced in the upcoming 16-week production period. But the question is this: What production mix would maximize her profits? Winnie has a few rules of thumb that she applies to this problem. First, she believes in product diversity, so at least 5,000 units of each kind of grill must be produced. In keeping with that rule, she does not want to overemphasize any one product, so no more than 20,000 of each kind of grill are produced. Of course, additional production limits are imposed by labor availability and by the number of components available.

Winnie assumes production will be done, as it has in the past, at the Wombat plant. Winnie calls that plan the Base Case. Recently, the president of one of Wombat's competitors called to say that she would have some excess capacity in the coming year and that her company would be willing to produce grills according to Wombat specifications. The grills would be made by the competitor, shipped to Wombat, packaged in Wombat's box, and sold by Wombat. That option would, in effect, increase Wombat's production capacity via outsourcing. Winnie calls that production planning option the Extension Case.

Winifred Pugh has called you in because she heard that you know how to use the Excel Solver to help with production planning problems. You must model the two different situations: the Base Case and the alternative Extension Case. After you have finished the Base Case, you will modify its spreadsheet to create the Extension Case. Comparing the results of the two cases will let Winnie play "what-if" scenarios with the production planning decision.

ASSIGNMENT 1: CREATING A SPREADSHEET FOR DECISION SUPPORT

In this assignment, you will produce a spreadsheet that models the business decision. In Assignment 1A, you will make a Solver spreadsheet to model the Base Case. In Assignment 1B, you will make a Solver spreadsheet to model the Extension Case. In Assignment 2, you will use the spreadsheet models to develop the information you need to recommend the best production planning mix. Then you will document your recommendation in a memo to Winnie Pugh. In Assignment 3, you will give your recommendation in an oral presentation.

First, you will create the spreadsheet models of the decision. Your spreadsheets should have the following sections:

- CHANGING CELLS
- CONSTANTS
- CALCULATIONS
- INCOME STATEMENT

Your spreadsheets also will include the decision constraints.

A Base Case spreadsheet skeleton is available to you, so you need not type it in. To access the spreadsheet skeleton, go to your Data files, select Case 9, and then select GRILL1.xls.

Assignment 1A: Creating the Spreadsheet—Base Case

You will model the production planning problem described. Your model, when run, will tell Winifred Pugh how many of each of the three kinds of grills to make in the upcoming 16-week production period.

A discussion of each spreadsheet section follows. For each section, you will learn (1) how the section should be set up and (2) what the logic of the sections' cell formulas is.

CHANGING CELLS Section

Your spreadsheet should have the changing cells shown in Figure 9-3.

	A	B
1	**GRILL MANUFACTURING SCHEDULE**	
2		BASE
3	**CHANGING CELLS**	**CASE**
4	CHARCOAL GRILL	1
5	BASIC LP GRILL	1
6	DELUXE LP GRILL	1

FIGURE 9-3   CHANGING CELLS section

You are asking the Solver model to compute the number of each kind of grill to make. Start with a 1 in each cell. The Solver will change each 1 as it computes the answer. Note that Winnie, obviously, would not want to be told to make a fraction of a grill.

CONSTANTS Section

Your spreadsheet should have the constants shown in Figures 9-4, 9-5, and 9-6. An explanation of the line items in each of those figures follows the figure.

	A	B	C	D
11	**CONSTANTS**			
12	BILL OF MATERIALS:	CHARCOAL	BASIC LP	DELUXE LP
13	CART FRAME	1	1	1
14	HOOD	1	1	1
15	LP BURNER	0	1	2
16	AUXILIARY ITEM PACK	0	1	2
17	VALVE REGULATOR	0	1	2
18	CONTAINER	1	1	1
19	CARDBOARD BOX	1	1	1

FIGURE 9-4   CONSTANTS section

The bill of materials is shown in Figure 9-4. The figure shows the components of each kind of grill and the number of each component required for each kind of grill. The Constants are further described in Figure 9-5.

	A	B
20	NUMBER OF COMPONENT ON HAND:	--
21	CART FRAME	50000
22	HOOD	50000
23	LP BURNER	40000
24	AUXILIARY ITEM PACK	40000
25	VALVE REGULATOR	40000
26	CONTAINER	50000
27	CARDBOARD BOX	50000
28	COMPONENT COSTS:	--
29	CART FRAME	50
30	HOOD	20
31	LP BURNER	75
32	AUXILIARY ITEM PACK	35
33	VALVE REGULATOR	30
34	CONTAINER	50
35	CARDBOARD BOX	2

FIGURE 9-5   CONSTANTS section, continued

- NUMBER OF COMPONENTS ON HAND: The available number of each component that can be used in production is shown. These numbers limit production at the Wombat plant; for example, because only 40,000 LP burners are on hand for making basic and deluxe grills, only 40,000 gas grills can be made.
- COMPONENT COSTS: The unit cost of each kind of component is shown.

The Constants are further described in Figure 9-6.

- SELLING PRICES: The selling price of each kind of grill is shown.
- LABOR REQUIRED PER UNIT (MINS): The labor minutes required to package each kind of unit are shown.

	A	B
36	SELLING PRICES:	--
37	CHARCOAL GRILL	300
38	BASIC LP GRILL	600
39	DELUXE LP GRILL	800
40	LABOR REQUIRED PER UNIT (MINS):	--
41	CHARCOAL GRILL	4
42	BASIC LP GRILL	9
43	DELUXE LP GRILL	11
44	NUMBER OF EMPLOYEES	50
45	HOURS PER WEEK	40
46	HOURS PER DAY	8
47	WEEKS AVAILABLE	16
48	WAGE PER HOUR	10
49	ADMIN COST PER UNIT	1
50	TAX RATE	0.3

FIGURE 9-6   CONSTANTS section, continued

- NUMBER OF EMPLOYEES: Fifty employees have been hired for the upcoming production period.
- HOURS PER WEEK: Employees will work 40 hours per week.
- HOURS PER DAY: Employees will work eight hours a day—no overtime is expected.
- WEEKS AVAILABLE: The production period will be 16 weeks.
- WAGE PER HOUR: Each employee will earn $10 an hour.
- ADMIN COST PER UNIT: A dollar administrative cost is incurred for each grill produced.
- TAX RATE: The tax rate on positive income before taxes is 30 percent. No taxes are paid on negative or zero net income before taxes.

CALCULATIONS Section

Your spreadsheet should calculate the values shown in Figure 9-7 and Figure 9-8. Calculated values are based on changing cell values, on constants, and/or on other calculated values; they will be used in the INCOME STATEMENT section or in the CONSTRAINTS. An explanation of the line items in each figure follows the figure.

	A	B	C	D	E
55	CALCULATIONS				
56	COMPONENTS USED IN MANUFACTURE:	CHARCOAL	BASIC LP	DELUXE LP	TOTAL
57	CART FRAME				
58	HOOD				
59	LP BURNER				
60	AUXILIARY ITEM PACK				
61	VALVE REGULATOR				
62	CONTAINER				
63	CARDBOARD BOX				

FIGURE 9-7   CALCULATIONS section

The number of COMPONENTS USED IN MANUFACTURE is a function of the number of each kind of grill that has been made (shown in the changing cells) and the number of the component used in each kind of grill (shown in the constants). The total of each component used is the sum of the components used in all kinds of grills. Calculations are continued in Figure 9-8.

	A	B
64	COST OF COMPONENTS USED:	--
65	CART FRAME	
66	HOOD	
67	LP BURNER	
68	AUXILIARY ITEM PACK	
69	VALVE REGULATOR	
70	CONTAINER	
71	CARDBOARD BOX	
72	LABOR MINUTES AVAILABLE	
73	LABOR MINUTES USED	
74	LABOR COSTS INCURRED	
75	ADMIN COSTS INCURRED	
76	REVENUE:	--
77	CHARCOAL GRILL	
78	BASIC LP GRILL	
79	DELUXE LP GRILL	
80	NUMBER OF GRILLS MADE	

FIGURE 9-8 CALCULATIONS section, continued

- COST OF COMPONENTS USED: The total cost of each component used in production is a function of the number of components used (a calculation) and the unit cost of the component (a constant).
- LABOR MINUTES AVAILABLE: This is a function of the number of production weeks, the hours per week and per day, and the number of workers.
- LABOR MINUTES USED: This is a function of the number of each kind of grill made (shown in the changing cells) and the minutes of assembly required per grill (a constant).
- LABOR COSTS INCURRED: This is a function of the number of hours used in production and the hourly pay rate.
- ADMIN COSTS INCURRED: This is a function of the number of grills made and the administrative cost per grill (a constant).
- REVENUE: The revenue for each type of grill is a function of the number made and sold (shown in the changing cells) and the selling price (a constant).
- NUMBER OF GRILLS MADE: This is the total of charcoal, basic LP, and deluxe LP grills made and sold.

INCOME STATEMENT Section

Your spreadsheet should be set up to compute the Wombat net income, as shown in Figure 9-9. An explanation of the line items follows the figure.

	A	B
87	INCOME STATEMENT	
88	REVENUE	
89	COST OF COMPONENTS	
90	LABOR COSTS	
91	ADMIN COSTS	
92	TOTAL COSTS	
93	NET INCOME BEFORE TAXES	
94	INCOME TAX EXPENSE	
95	NET INCOME AFTER TAXES	

FIGURE 9-9 INCOME STATEMENT section

- REVENUE: This is the sum of revenues calculated for each kind of grill.
- COST OF COMPONENTS: This is the total cost of all components used in production.
- LABOR COSTS: This amount is calculated elsewhere and can be echoed here.
- ADMIN COSTS: This amount is calculated elsewhere and can be echoed here.
- TOTAL COSTS: This is the sum of the cost of components, labor, and administrative costs.
- NET INCOME BEFORE TAXES: This is the difference between Revenue and Total Costs.
- INCOME TAX EXPENSE: Taxes are zero when Net Income Before Taxes is zero or negative. Otherwise, apply the tax rate (a constant) to the Net Income Before Taxes to compute the income tax expense.
- NET INCOME AFTER TAXES: This is the difference between Net Income Before Taxes and the Income Tax Expense.

Determine the constraints. Enter the Base Case decision constraints using the Solver. Run the Solver; when the Solver says that a solution has been found that satisfies the constraints, ask for the Answer Report.

When you are done, print the entire workbook, including the Answer Report sheet. Save the workbook one last time (Office Button—Save). Then to prepare for the Extension Case, use Office Button—Save As to make a new spreadsheet file. (GRILL2.xls would be a good name.)

Assignment 1B: Creating the Spreadsheet—Extension Case

A competing firm, Webster Grill Company, has a large manufacturing facility about 50 miles away from the Wombat plant. Webster says that it will have excess production capacity in the upcoming 16-week period, and it offers to make up to 15,000 grills for Wombat according to Wombat specifications.

The terms of the deal are that Webster would purchase the components from Wombat's source (i.e., Wombat would not have to ship any of its components to Webster). The finished grills would be shipped to Wombat unboxed. Wombat would inspect the grills, put them in boxes, and store them in its warehouse. The units would then be sold in the same way as the ones that Wombat makes—and for the same price.

Webster does not want to enter into this deal for a trivial number of grills. It will do this only if Wombat orders 2,000 of each kind of grill. Webster says that Wombat will have to pay the company $150 for a charcoal grill, $290 for a basic LP grill, and $430 for a deluxe LP grill.

Because this outsourcing option might result in higher net income for Wombat, Winnie wants to look at it closely. To take advantage of the option, she would need to order a specific number of charcoal, basic LP, and deluxe LP grills from Webster. Here are some factors that would change if Wombat were to enter into this deal:

- The administrative cost of the Webster-produced units would increase to $5 per unit due to the cost of transportation and inspection. (The administrative cost of Wombat-produced grills would remain the same.)
- Wombat would order at least 2,000 charcoal grills, 2,000 basic LP grills, and 2,000 deluxe LP grills. Wombat would order no more than 15,000 grills in total from Webster.
- Wombat would plan on selling no more than 25,000 of each kind of grill (up from 20,000 in the Base Case). This means that the number of Wombat-produced charcoal grills plus the number of Webster-produced charcoal grills would not exceed 25,000.

Modify the Extension Case spreadsheet and related constraints to reflect the outsourcing option. You should modify the changing cells as shown in Figure 9-10.

	A	B
1	**GRILL MANUFACTURING SCHEDULE**	
2		EXTENSION
3	CHANGING CELLS	CASE
4	CHARCOAL GRILL-MAKE	1
5	BASIC LP GRILL-MAKE	1
6	DELUXE LP GRILL-MAKE	1
7	CHARCOAL GRILL-OUTSOURCED	1
8	BASIC LP GRILL-OUTSOURCED	1
9	DELUXE LP GRILL-OUTSOURCED	1

FIGURE 9-10   Extension Case CHANGING CELLS section

Include Webster's costs per unit in the Constants and compute the cost of Webster-produced grills in the Calculations. The total cost of outsourced grills should be included as a line item in the Extension Case Income Statement.

Run the Solver. When the Solver says that a solution has been found that satisfies the constraints, ask for an Answer Report. When you are done, print the entire workbook, including the Solver Answer Report. Then save the file, close the file, and exit Excel.

ASSIGNMENT 2: USING THE SPREADSHEET FOR DECISION SUPPORT

You have built the Base Case and the Extension Case models because you want to know the profitability of each production planning approach. Now you will complete your work by (1) using the worksheets and Answer Reports to gather data and (2) documenting your analysis and recommendation in a memo and, if your instructor specifies, in an oral presentation.

Assignment 2A: Using the Spreadsheets to Gather Data

Since you printed the spreadsheet and Answer Report for each case, you can see the results of each investment approach. Winnie Pugh wants to know how net income after taxes differs between the two alternatives. If net income is significantly higher in the Extension Case, she will take Webster's offer, despite the quality-control risks. Otherwise, she will not outsource.

You should summarize the key data in a table that will be included in your memo. The form of that table is shown in Figure 9-11. (In the Base Case, there is no outsourcing of grills, so Webster-produced grills are NA.)

	Base Case	Extension Case
Charcoal grills—Wombat-produced		
Basic LP grills—Wombat-produced		
Deluxe LP grills—Wombat-produced		
Charcoal grills—Webster-produced	NA	
Basic LP grills—Webster-produced	NA	
Deluxe LP grills—Webster-produced	NA	
Total labor minutes used—Wombat-production		
Net income after taxes		

FIGURE 9-11   Format of table to insert into memo

Number of grills made can be found in the Answer Reports and in the worksheet changing cells. Net income after taxes can be found in the Answer Reports and in the worksheets. Total labor minutes used in Wombat production can be found in the Calculations areas of the worksheets.

Assignment 2B: Documenting Your Recommendation in a Memo

Use Word to write a brief memo to Winifred Pugh about the results of your analysis. Observe the following requirements:

- Include a proper heading (TO/FROM/DATE/SUBJECT). You may use a Word memo template (Click **Office Button**, click **New**, click the **Memo** template in the Templates section, choose Contemporary Memo, then click **OK**. The first time you do this you may need to click **Download** to install the template).
- Briefly state the problem and the decision to be made, but do not provide much background—you can assume Winnie is aware of the problem. Briefly state your analytical method and provide the results of the analysis.
- Give Winnie Pugh your recommendation, which should be based on the highest net income after taxes. Explain your reasoning.
- Support the recommendation graphically by including a summary table in your memo, as shown in Figure 9-11. Create the table by following the procedure described next.

Enter a table into the Word memo using the following procedure:

1. Click the cursor where you want the table to appear in the document.
2. In the Insert group, select the Insert drop-down menu.
3. Select Insert Table.

4. Choose the number of rows and columns.
5. Click **OK**.

ASSIGNMENT 3: GIVING AN ORAL PRESENTATION

Your instructor may request that you present your analysis and recommendations in an oral presentation. If so, assume Winnie Pugh is very impressed by your analysis and she wants you to give a presentation to some of her managers. Prepare to explain your analysis and recommendation to the group in ten minutes or less. Use visual aids or handouts that you think are appropriate. (Tutorial F has information on how to prepare and give an oral presentation.)

DELIVERABLES

Assemble the following deliverables for your instructor:

1. Printout of your memo
2. Spreadsheet printouts
3. Disk that contains your memo and spreadsheet files. Do not provide a CD, as it would be read-only.

Staple the printouts together, placing the memo on top.

PART 4

DECISION SUPPORT CASES
USING BASIC EXCEL FUNCTIONALITY

THE BALLPARK DECISION

Decision Support Using Excel

PREVIEW

The owner of your city's major league baseball team wants a new stadium and a new operating agreement. The owner offers to cover the expenses of building the new stadium; but in return, she wants the city to give up various future operating revenues. In this case, you will use Excel to help the city evaluate the owner's offer.

PREPARATION

Before attempting this case, you should:

- Review spreadsheet concepts discussed in class and/or in your textbook.
- Complete any exercises that your instructor assigns.
- Complete any parts of Tutorials C and D that your instructor assigns or refer to them as necessary.
- Review file-saving procedures for Windows programs, which are discussed in Tutorial C.
- Refer to Tutorial F as necessary.

BACKGROUND

Your city has a major league baseball team. The team plays in a baseball stadium that was built by the city 40 years ago. The baseball stadium is owned by the city, and each year the team leases the ballpark from the city for its home games.

Under the lease agreement, the city keeps parking and concession receipts. (Concession receipts are proceeds from the sale of items such as hot dogs and souvenirs.) The team keeps ticket sales but pays a tax to the city on each ticket sold.

The old ballpark is rather antiquated, especially when compared to other newer major league parks. Everyone involved with the baseball team and most city government officials agree that the old ballpark must be replaced soon.

The owner of the team wants not only a new baseball stadium but also a new operating agreement with the city. She wants to keep all or part of the parking and concession revenues, and she wants other benefits as well. In fact, if she does not get a better deal, she has threatened to move the team to another city.

Her claim—which has some validity—is that with a better venue, she can make more money, which will let her put together a better team, one that has a better chance to win a championship for the city. The city must take the owner's threat seriously, because a number of other cities have, in recent years, lost their baseball teams when they refused to meet the same kinds of demands.

The owner proposes the following deal to the city:

- The team owner will buy land in the city's downtown area and build a stadium, which the team owner will own. Expanded parking will be a feature of the new baseball stadium. The cost of land and construction is estimated to be a half billion dollars. Construction should be done in 2007, and the new stadium will be ready for baseball in 2008. Thus, the city's taxpayers would not spend a penny, but there would be a new stadium within the city limits.
- Whenever anyone puts up a building in town, the person must pay a tax on the construction materials used. The team owner would not have to pay this one-time construction cost tax.
- A city property tax funds public schools in the city. The tax is usually paid on a baseball stadium's land and fixtures. The team wants to be exempt from paying this yearly tax on the new baseball stadium.

- Currently, the team pays a 5 percent tax to the city on baseball ticket sales. For example, on a $20 seat, the team pays $1 to the city. Once the new ballpark is built, the team would not have to pay the ticket sales tax to the city.
- Currently, as owner of the stadium, the city keeps concession sales and automobile parking revenues. The team wants to keep a healthy portion of these receipts—in fact, the team wants to keep all of this money.

In addition to those new terms, the city must consider another cost it will have to bear. Currently, many fans take the subway and buses to games. With the new stadium's expanded parking, more people will likely choose to drive to the stadium, which would reduce the city's subway and bus receipts.

To summarize, under the owner's proposed deal, the city would get something right away for little cost—a new stadium costing taxpayers nothing. In addition, the city could eventually get a better baseball team. But there would be a price to pay. In future years, the city would have to give up various revenues. In effect, the new deal is structured as follows: the owner is proposing to lend the city a great deal of money today; and in exchange, the city will repay the loan in future years by letting the owner keep the revenue that the city would have otherwise collected.

The team owner and the mayor are starting to negotiate the team's offer. In recent years, this sort of negotiation has become commonplace in the American sports economy. A team owner threatens to move to a different city if the team is not given a new place to play or a better operating agreement. But the team owner does not really want to move, because nobody knows whether a better deal is available elsewhere. The city's political leaders want to help, but they don't want to be embarrassed. They don't want voters to be unhappy that a popular team left town on their watch. On the other hand, they don't want to be accused of giving money away to a rich team owner. The city could entice some other team to move to town, but a new deal might be worse than the one they could have negotiated with the existing team.

The mayor of your city knows that the team owner wants to stay, but also knows she probably would leave if a new baseball stadium is not built. The mayor also realizes that the city—and the team—do in fact need a new baseball stadium. To further complicate matters, the decision has attracted much media interest. The question that dominates local headlines and talk radio shows is whether the city will give up too much to get a ballpark.

The new baseball stadium would be built in 2007 and would open for play in 2008. There's no telling when the team and the city will want yet another new baseball stadium after the new one, but 20 years is a good guess. So the time horizon would be the next 20 baseball seasons, in other words, 2008–2027. In that time, assume the field would be used only for baseball.

The mayor needs to think about the economics of the team's offer from the standpoint of the city's interests. You have been called in to make a spreadsheet that will help the mayor understand the trade-offs involved in this decision. The mayor tells you that the baseball team has, in effect, offered to give the city a half billion dollars in the form of a new ballpark. "But in sports," the mayor adds, "there's no such thing as a gift." Indeed, the "gift" of the baseball stadium would be paid back by the city in the form of lost future revenue from tax receipts, concession sales, etc. The mayor wants to know what the team's proposal would cost the city's taxpayers the next 20 years—that is the way the media and the public understand the question, and that is the way it will have to be answered.

ASSIGNMENT 1: CREATING A SPREADSHEET FOR DECISION SUPPORT

In this assignment, you will produce a spreadsheet that models the business problem. Then in Assignment 2, you will write a memo to the mayor that explains your findings and recommends a course of action. In addition, in Assignment 3, you may be asked to prepare an oral presentation of your analysis and recommendation.

First, you create the spreadsheet model of the proposal. The model is for the 20 years of the new operating agreement, 2008–2027. You will be given instructions on how to set up each spreadsheet section before you enter cell formulas. Your spreadsheet should include these sections:

- CONSTANTS
- INPUTS
- SUMMARY OF KEY RESULTS
- CALCULATIONS

- COSTS BORNE BY THE CITY
- PRESENT VALUE OF COSTS BORNE BY THE CITY

A discussion of each section follows. The spreadsheet skeleton is available to you, so you need not type it in. To access the spreadsheet skeleton, go to your Data files. Select Case 10 and then select STADIUM.xls.

CONSTANTS Section

Your spreadsheet should have the constants shown in Figure 10-1. Those constants will be used in the spreadsheet formulas. (*Note:* Only the first few years of the analysis period are pictured in Figure 10-1; the actual spreadsheet columns extend to depict 20 years.) An explanation of the line items follows the figure.

	A	B	C	D	E	F
1	**HOW MUCH DOES THAT BASEBALL STADIUM COST?**					
2						
3	CONSTANTS	2007	2008	2009	2010	2011
4	TAX ON TICKETS	0.05	0.05	0.05	0.06	0.06
5	CONCESSION REVENUE PER ATTENDEE	10	10	10	11	11
6	CONSTRUCTION COST -- MATERIALS	250000000	NA	NA	NA	NA
7	CONSTRUCTION COST -- OTHER	250000000	NA	NA	NA	NA
8	TAX RATE ON CONSTRUCTION MATERIALS	0.02	NA	NA	NA	NA
9	NUMBER OF AUTOS PER GAME	5000	5000	5000	5000	5000
10	COST TO PARK AUTO	10	10	10	11	11
11	SUBWAY AND BUS REVENUE PER GAME	10000	10100	10200	10300	10400

FIGURE 10-1 CONSTANTS section

- TAX ON TICKETS: Currently, the team pays 5 percent of ticket sales to the city. The rate is expected to go up in future years. If a new stadium is built, the team wants to be exempt from paying this tax.
- CONCESSION REVENUE PER ATTENDEE: On average, fans spend $10 per game on refreshments and souvenirs. This amount is expected to increase in future years.
- CONSTRUCTION COST—MATERIALS: The cost of the new stadium's construction materials will be a quarter billion dollars, to be spent in 2007.
- CONSTRUCTION COST—OTHER: The cost of the land, parking lot, and other fixtures will be a quarter billion dollars, which will be spent in 2007.
- TAX RATE ON CONSTRUCTION MATERIALS: This rate is 2 percent.
- NUMBER OF AUTOS PER GAME: The new baseball stadium will have expanded parking, and it is expected that 5,000 cars will park at a game, on average.
- COST TO PARK AUTO: The cost to park for a game will be $10. This rate is expected to go up in future years.
- SUBWAY AND BUS REVENUE PER GAME: Currently, the city earns, on average, $10,000 a game from fans taking the subway and buses. Some of this amount will be lost under the team's proposal due to the expanded parking. (The percentage lost will be revealed in the Calculations section.)

INPUTS Section

Your worksheet should have the four inputs shown in Figure 10-2, which will be used by the spreadsheet's formulas. Your instructor may tell you to apply Conditional Formatting to the input cells so that out-of-bounds values are highlighted in some way. An explanation of the line items follows the figure.

	A	B	C	D	E	F
13	INPUTS	ALL YEARS	2008-2012	2013-2017	2018-2022	2023-2027
14	INFLATION RATE (EACH YEAR)		NA	NA	NA	NA
15	CITY SHARE OF PARKING & CONCESSIONS		NA	NA	NA	NA
16	TEAM AVOIDS PAYING PROPERTY TAX? (Y/N)		NA	NA	NA	NA
17	PLAYOFFS? (P = YES, L = LOSERS)	NA				

FIGURE 10-2 INPUTS section

- INFLATION RATE (EACH YEAR): Over the course of the 20-year period, the inflation rate will have an effect on the estimated cost of tickets and on other factors. For each year, the user enters an expected amount for the inflation rate. For example, if the rate is expected to be 5 percent, the user would enter .05.
- CITY SHARE OF PARKING & CONCESSIONS: The user enters a number between 0 and 1.0. Zero would indicate that the team keeps all of the receipts from parking and concession sales. An entry of 1.00 would indicate that the city keeps all of those receipts. An entry of .40 would indicate that the team keeps 60 percent and the city keeps 40 percent, in which case, the city would be giving up 60 percent of the receipts. The value entered applies to all years. The team owner and the city would negotiate this value.
- TEAM AVOIDS PAYING PROPERTY TAX? (Y/N): The team does not want to pay property tax on the new stadium. In this city, property taxes finance the public schools; so giving up the tax could be controversial for the mayor. Here the spreadsheet user enters *YES* if the team will avoid paying the property tax (i.e., not pay the tax). The user enters *NO* if the team will pay the tax.
- PLAYOFFS? (P = YES, L = LOSERS): The user enters *P* if the team is expected to make the playoffs. The user enters *L* if the team is not expected to make the playoffs. For this purpose, the 20 years has been divided into four 5-year periods. The letter entered in each of the five-year periods indicates the team's prospects for success or failure for those five baseball seasons. An entry of PLLL in the four input cells would mean that the user thinks the team will make the playoffs in all of the years 2008–2012, but then not make the playoffs at all in the years 2013–2027.

SUMMARY OF KEY RESULTS Section

Your worksheet should have the two key results shown in Figure 10-3. Those values are echoed from elsewhere in your spreadsheet. An explanation of the line items follows the figure.

	A	B	C	D	E	F
19	SUMMARY OF KEY RESULTS		ALL YEARS			
20	TOTAL COST TO THE PEOPLE	NA		NA	NA	NA
21	PRESENT VALUE OF TOTAL COST TO PEOPLE	NA		NA	NA	NA

FIGURE 10-3   SUMMARY OF KEY RESULTS section

- TOTAL COST TO THE PEOPLE: The mayor wants to know the total cost of the team's proposal that would be borne by the city's taxpaying population. This value corresponds to the revenues given up by the city in the 20-year period of the agreement.
- PRESENT VALUE OF TOTAL COST TO PEOPLE: This is the present value of the revenues foregone in the upcoming 20 years. (The concept of present value will be explained in a later section.) The calculation, in effect, expresses the total sum of the values for the next 20 years as one current-year value, and it is to be used for comparison with the half-billion-dollar cost of the stadium.

CALCULATIONS Section

This section calculates the values shown in Figure 10-4. Values are calculated by formula. The values are then used in other calculations and/or in the spreadsheet body that follows. An explanation of the line items follows the figure.

	A	B	C	D	E	F
23	CALCULATIONS	2007	2008	2009	2010	2011
24	NUMBER OF HOME GAMES	NA				
25	AVERAGE COST OF TICKET	20.00				
26	AVERAGE ATTENDANCE PER GAME	NA				
27	REVENUE FROM SELLING TICKETS	NA				
28	TICKET TAX REVENUE	NA				
29	TOTAL CONCESSION REVENUE	NA				
30	TOTAL PARKING REVENUE	NA				
31	PROPERTY TAX ON STADIUM	1000000				
32	TAXES ON CONSTRUCTION MATERIALS	NA		NA	NA	NA
33	SUBWAY, BUS REVENUE LOST (@ .33)	NA				

FIGURE 10-4   CALCULATIONS section

- NUMBER OF HOME GAMES: In a given season, if the team does not make the playoffs, there will be 81 home games. If the team does make the playoffs, the number of home games depends on how well the team continues to perform—a factor that is difficult to estimate. If a first-round exit occurs, there would be only two or three extra games. But if the team played a seven-game World Series, as many as ten extra games would be played at home. For the purpose of this model, assume if the team makes the playoffs that there will be five extra home games.
- AVERAGE COST OF TICKET: The average cost of a single ticket in 2007 will be $20. It is assumed that ticket prices will increase by one-half the rate of inflation from year to year. Example: Suppose inflation is 10 percent. The price in 2008 would then be $20*1.05, or $21. The price in 2009 would be $21*1.05, continuing the same way for all years. But the owner and the city agree that in no year will the average ticket price exceed $50, regardless of the inflation rate. *Hint*: Use an IF() statement here or use the MIN() function.
- AVERAGE ATTENDANCE PER GAME: In periods when the team is expected to make the playoffs, average attendance per game would be 45,000. Otherwise, attendance is expected to be 30,000 per game. The number of games in a season has already been calculated. (*Hint:* Expected performance is shown in the inputs in five-year blocks).
- REVENUE FROM SELLING TICKETS: This is ticket revenue for the season. It is a function of the average ticket price, the attendance per game, and the number of games in the season.
- TICKET TAX REVENUE: This is the tax imposed on the season revenue expected from ticket sales. It is a function of the tax rate and the total ticket revenue for the season.
- TOTAL CONCESSION REVENUE: This is a function of the year's attendance and the average amount fans spend on concessions per game.
- TOTAL PARKING REVENUE: This is a function of the average number of cars at a game, the number of games, and the average spent on parking per game.
- PROPERTY TAX ON STADIUM: The city estimates that property tax on the new stadium would be $1 million in 2007, with the tax increasing at the rate of inflation each year thereafter.
- TAXES ON CONSTRUCTION MATERIALS: This amount would be paid in 2008 only. It is a function of the cost of construction materials and of the construction cost tax rate. Both factors are constants.
- SUBWAY, BUS REVENUE LOST (@.33): This is a function of the number of games and the subway and bus revenue per game. The city assumes one-third of this revenue would be lost due to expanded stadium parking.

COSTS BORNE BY THE CITY'S PEOPLE Section

This section is the spreadsheet body. Shown in Figure 10-5, this section summarizes and totals the revenue that the city would forego under the team's proposal. An explanation of the line items follows the figure.

	A	B	C	D	E	F
35	COSTS BORNE BY THE CITY'S PEOPLE	2007	2008	2009	2010	2011
36	PROPERTY TAXES FOREGONE	NA				
37	SUBWAY, BUS REVENUE LOST (@ .33)	NA				
38	FOREGONE TICKET TAX	NA				
39	FOREGONE CONCESSION REVENUE	NA				
40	FOREGONE PARKING REVENUE	NA				
41	CONSTRUCTION MATERIAL TAX FOREGONE	NA		NA	NA	NA
42						
43	COSTS BORNE BY PEOPLE EACH YEAR	NA				
44	TOTAL COSTS BORNE BY PEOPLE	NA		NA	NA	NA

FIGURE 10-5   COSTS BORNE BY THE CITY'S PEOPLE section

- PROPERTY TAXES FOREGONE: If the team does not pay the property tax (an input in Figure 10-2), the taxpayers will not collect the property taxes (a calculation).
- SUBWAY, BUS REVENUE LOST (@.33): This amount has been calculated elsewhere and should be echoed here.
- FOREGONE TICKET TAX: This amount has been calculated elsewhere and should be echoed here.
- FOREGONE CONCESSION REVENUE: Total concession revenue has been calculated elsewhere. The share of this amount that the city will *give up* should be computed here.

- **FOREGONE PARKING REVENUE:** Total parking revenue has been calculated elsewhere. The share of this amount that the city will *give up* should be computed here.
- **CONSTRUCTION MATERIAL TAX FOREGONE:** This amount has been calculated elsewhere and should be echoed here.
- **COSTS BORNE BY PEOPLE EACH YEAR:** This is the sum of the six costs borne by the city's people during the year.
- **TOTAL COSTS BORNE BY PEOPLE:** This is the sum of all costs borne in the 20-year period of the operating agreement. (Once calculated here, this amount is echoed to the SUMMARY OF KEY RESULTS section.)

PRESENT VALUE OF COSTS BORNE BY THE CITY'S PEOPLE Section

This section continues the spreadsheet body. The calculation of present value is best explained with an example.

As discussed earlier, the operating agreement is structured such that the city would, in effect, "pay back" the half billion dollars the team owner spent on construction of the new stadium, as though it were a loan. The city would "repay" the money in 20 unequal installments. Each installment would correspond to an amount of future revenue that the city would normally collect but that it would turn over to the owner instead. Is this 20-year deal worthwhile to the city?

One of the complicating factors in answering that question is that a dollar paid tomorrow does not have the same value as a dollar paid today. While that difference may not be as dramatic as the difference between apples and oranges, it is considerable enough that you must perform a conversion before you can adequately compare tomorrow's dollar with today's. This conversion, which accounts for changes due to the interest rate at which today's dollar might be invested, involves restating the future cash flows in terms of current year dollars through the use of a discount factor (sometimes called a *discount rate*). Suppose a person could borrow or invest money at 10 percent and amounts would be discounted at that rate.

Here is a simplified example. Suppose someone lends you $15,000 in year 0, then requires you to repay $6,000 a year later, $7,000 a year after that, and $8,000 a year after that. Your discount rate is 10 percent. You see immediately that the repayments total $21,000, but you know that comparing that amount to your principal of $15,000 is not the best way to evaluate the loan because it doesn't account for the present value of your future cash flows. Should you agree to that loan arrangement? Figure 10-6 features a computation that gives the present value of each of your future repayments.

Period (years)	Annual Payment	Discount Factor at 10%	Present Value of Cash Flow at 10%
1	$ 6,000	$1/(1 + .10)^1 = .909$	$ 5,454
2	7,000	$1/(1 + .10)^2 = .826$	5,782
3	8,000	$1/(1 + .10)^3 = .751$	6,008
Total	$21,000		$17,244

FIGURE 10-6 Computation of present value of repayment

As Figure 10-6 shows, $6,000 paid back at the end of the first year is the same as paying back $5,454 at the outset. Paying back $7,000 at the end of the second year is the same as paying back $5,782 at the outset; and finally, $8,000 paid back at the end of the third year is equivalent to paying $6,008 today. Now you can see that accepting this loan is, in fact, not a good idea. You get $15,000 today, but you must repay more than that in today's dollars—$17,244 in total.

In the case of the ballpark, recall that the city gets an immediate benefit of a half billion dollars and then pays for that benefit over 20 years. To assess whether that "repayment" scheme is beneficial to the city, you need to compute the present value of the amounts of future revenue that the city would forego under the team's proposal. That computation is shown in Figure 10-7. An explanation of the items follows the figure.

	A	B	C	D	E	F
46	**PRESENT VALUE OF COSTS BORNE BY THE CITY'S PEOPLE**					
47				PRESENT		
48				VALUE OF		
49		COST BORNE	DISCOUNT	COST BORNE		
50	YEAR	IN YEAR	FACTOR	IN YEAR		
51	2008					
52	2009					
53	2010					
54	2011					
55	2012					
56	2013					
57	2014					
58	2015					
59	2016					
60	2017					
61	2018					
62	2019					
63	2020					
64	2021					
65	2022					
66	2023					
67	2024					
68	2025					
69	2026					
70	2027					
71						
72	ALL YEARS		NA			

FIGURE 10-7   PRESENT VALUE OF COSTS BORNE BY THE CITY'S PEOPLE section

- COST BORNE IN YEAR: These values, which are composed of foregone revenues, are echoed from another part of the spreadsheet.
- DISCOUNT FACTOR: Here you calculate each year's discount factor using the formula shown in Figure 10-6's Discount Factor at 10% column. For the discount rate for each year, use that year's inflation rate, which was an input in Figure 10-2.
- PRESENT VALUE OF COST BORNE IN YEAR: This is the product of costs borne in a given year and that year's discount factor.
- ALL YEARS: You should total the costs borne in all of the years and the present values of those costs.

ASSIGNMENT 2: USING THE SPREADSHEET FOR DECISION SUPPORT

Now you will complete the case by (1) using the spreadsheet to gather the data the mayor needs in order to decide whether the team owner's offer would make financial sense, (2) documenting your findings and recommendation in a memo, and (3) giving an oral presentation if your instructor specifies.

Assignment 2A: Using the Spreadsheet to Gather Data

You have built the spreadsheet to let the mayor play "what-if" scenarios with some of the decision variables.

The mayor knows a bit about finance. ("I have an MBA!" she once crowed at a city council meeting.) So she thinks that the half-billion-dollar construction cost should be compared with the present value of the costs borne by the people. She reasons that if the present value of those costs is around a half billion dollars (or less, of course), the deal would probably look good for the people of the city. But if the present value of the costs is much higher than a half billion dollars, it will be harder to make a case in support of the new stadium.

But the mayor also knows that most citizens do not understand discounting. So most people probably think that, say, a billion dollars of lost revenue—even if it were lost over 20 years—would be too high a price to pay for the new stadium. To help persuade the citizens and the media, the mayor thinks that today's half-billion-dollar construction cost benefit should be compared with the present value of the costs borne by the people later on.

The mayor doesn't want to waste time. She wants to begin negotiating with the team owner now. But she needs to know what is and is not financially acceptable. That is where your spreadsheet is needed.

The mayor envisions three possible scenarios. One of them should—she hopes—be favorable to the city (i.e., the owner would not like it). One is likely to be unfavorable to the city (i.e., the owner would love it). The third falls somewhere in the middle. Use your spreadsheet to develop the three scenarios.

Here are the mayor's guidelines for your "what-if" analysis.

- The mayor thinks that inflation could be 1 percent, 3 percent, or 5 percent.
- The mayor knows that the owner wants a good share of the concessions. Assume the city would be able to keep 50 percent, 42 percent, or 33 percent.
- For political reasons, the mayor would like to collect the property tax. But she has decided not to insist on this.
- The baseball team has not been very good lately. These two patterns are possible: LLPP and LPPP.

You should run your "what-if" analysis. You can use the Scenario Manager to organize the effort, but doing so is not required. You can simply enter values manually, writing the results on paper as you see them in the Summary of Key Results. In the end, summarize the results as shown in Figure 10-8. (*Note:* The headings in the figure reflect the mayor's assumptions about how favorable a scenario would be for the city.)

	Favorable	Middle	Unfavorable
Inflation rate	.05	.03	.01
City share	.50	.42	.33
Team avoids property tax?	NO	YES	YES
Playoff pattern	LLPP	LLPP	LLPP
Total costs borne by people			
Present value of total costs borne by people			

FIGURE 10-8 Summary of the three negotiating scenarios

When you have finished gathering the data, print the worksheet (with any set of inputs). Then save the spreadsheet one last time (Office Button—Save).

Assignment 2B: Documenting Your Findings and Recommendation in a Memo

Now you will document your findings in a memo. The mayor wants answers to these four questions:

1. What seems to be the most favorable scenario?
2. What seems to be the least favorable scenario?
3. Does it seem as if the city would be better off financially with high inflation? Does it seem as if the city would be better off with a losing baseball team? (To find out, check results with the playoff pattern LPPP.) If inflation and a losing team actually seem *beneficial*, should the mayor take into account other values in the spreadsheet?
4. Does the mayor need to know more, or is she ready to negotiate with the team owner? What is your overall recommendation to the mayor?

Follow these guidelines when preparing your memo in Word:

- Use a proper heading (TO/FROM/DATE/SUBJECT). You may use a Word memo template (Click **Office Button,** click **New**, click the **Memo** template in the Templates section, choose **Contemporary Memo,** then click **OK**. The first time you do this you may have to select **Download** to install the template).
- Briefly outline the situation. However, you need not provide much background—you can assume the mayor is familiar with the problem.
- Answer the mayor's four questions in the body of the memo.
- Support your claims graphically by including the table shown in Figure 10-8.

Enter the table in your memo, using the following procedure:

1. Click the cursor where you want the table to appear in the document.
2. In the **Insert** group, select the **Insert** drop-down menu.
3. Select **Insert Table**.
4. Choose the number of rows and columns.
5. Click **OK**.

ASSIGNMENT 3: GIVING AN ORAL PRESENTATION

Your instructor may request that you present your results in an oral presentation. If so, assume the mayor is impressed by your analysis and has asked you to give a presentation explaining your results to the city's management. Prepare to explain your results to the group in ten minutes or less. Use visual aids or handouts that you think are appropriate. (Tutorial F has information on how to prepare and give an oral presentation.)

DELIVERABLES

Assemble the following deliverables for your instructor:

1. Printouts of your memo
2. Spreadsheet printouts
3. Disk that contains your memo and your spreadsheet files. Do not provide a CD, as it would be read-only.

Staple the printouts together, placing the memo on top. If there is more than one .xlsx file on your disk, write your instructor a note stating the name of your model's .xlsx file.

PART 5

INTEGRATION CASE: USING ACCESS AND EXCEL

ANALYSIS OF A LOW INTEREST RATE LOAN PORTFOLIO

Decision Support Using Access and Excel

PREVIEW

A bank is now able to increase the interest rate on some loans it made five years ago. Raising the interest rate might increase interest revenue, but borrowers may have a more difficult time making their monthly payment to the bank. Should the bank refinance those loans? You have been called in to automate the analysis, using an Excel spreadsheet and Access database.

PREPARATION

Before attempting this case, you should:

- Review spreadsheet and database concepts discussed in class and/or in your textbook.
- Complete any exercises that your instructor assigns.
- Obtain the database file LOWLOANS.accdb from your Data files in the Case 11 folder.
- Review any part of Tutorials A, B, C, or D that your instructor specifies or refer to them as necessary.
- Review file-saving procedures for Windows programs, which are discussed in Tutorial C.
- Refer to Tutorial E as necessary for information on Excel data table processing and the VLOOKUP, CUMPRINC, and PMT functions.

BACKGROUND

The Teaser Loan Company is a bank that makes loans to people so they can buy their own homes. Suppose you want to buy a house that sells for $100,000 but you do not have $100,000. You do, however, have enough money to make a small down payment; for example, $5,000. You would contact Teaser Loan Company, and a loan representative would be assigned to you. The representative would assess your situation and decide if Teaser should lend you money to buy the house. If you qualified, Teaser would lend you the remainder of the money (in this case, $95,000). The lender requires you to use your new home as collateral on the loan, which means if you do not make your loan payments, the lender can repossess your house. In other words, the lender can take the house from you and sell it to satisfy the loan. That type of loan is often called a mortgage. You own the house; but because you still owe the amount you borrowed, you can lose the home if you do not pay off the loan.

Home loans are usually for 15 to 30 years and require the borrower to make monthly payments. Although interest rates on home loans can spike or dip significantly, the rates typically range from 6 percent to 8 percent. To appreciate how interest rates affect monthly payments, consider the following example: Assume a $100,000 house is purchased, with the borrower putting $1,000 down and borrowing $99,000. The $99,000 owed is called the loan principal. Suppose the loan is for 20 years—thereby involving 240 monthly payments. If the interest rate were 6 percent, each monthly payment would be $709. But if the interest rate were just 1 percent, each monthly payment would be only $455, a difference of $254.

A monthly loan payment is composed of two parts: (1) an amount to pay off part of the loan principal and (2) an amount to pay for interest on the remaining loan balance. Recall the $99,000 loan with an interest rate of 6 percent. The first month's $709 payment would pay off $214 of the loan balance (rounding for the pennies). The remaining loan principal balance would then be $99,000 less $214, or $98,786. The rest of the first month's payment ($495) would pay for the interest on the loan principal balance owed at the start of the month. The second $709 monthly payment would pay off $215 of the loan balance, with the rest ($494) going to interest on the now somewhat smaller remaining loan principal. Note that successive payments have smaller and smaller interest amounts, because the amount of principal owed declines with each payment.

Now recall the difference between the monthly payment at 6 percent ($709) and the monthly payment at 1 percent ($455). Some people could not afford to buy a house if they had to come up with $709 a month, but they might be able to buy if they were required to pay only $455 a month. That is where lenders such as Teaser Loan Company offer their services. Teaser will lend at very low rates for the early part of the loan, then convert the loan to a higher rate after a few years. The low rate at the start is called the teaser rate, because it entices the prospective home owner to borrow money and purchase a home.

The benefit to the borrower of such loans is that the prospective home buyer can purchase a home now. The challenge to the borrower of such loans is finding a way to make the monthly payments when the interest rate goes up and the payments increase.

Currently, Teaser has 100 twenty-year loans that it made almost 5 years ago at an interest rate of 1 percent. The loans were made for houses in Maryland, Pennsylvania, and Delaware. Under the terms of the loans and current laws, Teaser has the option, at the 5-year mark, to increase the loans up to the prime rate. The prime rate is the rate Teaser charges when it lends to its best customers. The rate is expected to be 6 percent at the 5-year mark. When a loan is converted to a new rate, Teaser, in effect, makes a new 15-year loan for the remaining loan principal balance at the new interest rate. Example: Assume Teaser made a $99,000 loan at 1 percent. Now suppose $14,000 in principal was paid off during the first 5 years of the loan. Teaser will rewrite the loan for $85,000 at the new rate. The new rate can be as high as the prime rate.

Some important points to keep in mind are the following: Teaser could opt to rewrite a loan for less than the prime rate. Also, Teaser is not obligated to rewrite any loans. Furthermore, it could rewrite some of the loans and leave others unchanged. Lastly, of those loans that are rewritten, some could be at the full prime interest rate and others could be at less than that rate.

At first, you might think that Teaser should rewrite all of its loans for the maximum rate allowed. Higher rates mean higher interest revenue, right? You should realize, however, that the increased monthly payment might force many borrowers into default, meaning that the borrower might not be able to make the higher monthly payment.

In default situations, the lender has difficult choices. Should the lender let the borrower make zero payments for a few months and hope the borrower's prospects will improve so the borrower can make the higher payment? Should the loan be rewritten yet again, but at a lower rate? Should the lender simply repossess the house and sell it for whatever amount it can get?

The likelihood that a borrower will default is a function of the relationship between the monthly loan payment to the borrower's monthly disposable income. (Monthly disposable income is the cash available to pay for housing, food, clothing, and other living expenses.) When the monthly loan payment to disposable income ratio is low, the borrower is likely to make the monthly payment. However, as the payment to income ratio increases (as it would if the interest rate on the loan were raised), the likelihood of default increases as well.

Teaser's management has developed a table, shown in Figure 11-1, that summarizes payment to income ratios and their corresponding likelihood of default.

Payment to Income Ratio	Default Likelihood	Payment to Income Ratio	Default Likelihood
.20	.01	.70	.29
.25	.02	.75	.36
.30	.03	.80	.44
.35	.04	.85	.53
.40	.05	.90	.63
.45	.06	.95	.74
.50	.09	1.00	.86
.55	.13	1.05	.97
.60	.18	1.10	.98
.65	.23	1.15	.99

FIGURE 11-1 Default likelihood table

Consider the following example: Suppose a borrower has a house payment of $500 a month and a disposable income of $1,000 a month, and thus a payment to income ratio of .50. According to Figure 11-1, that ratio indicates a 9 percent chance that the borrower will be unable to make the loan payments. If the same borrower's disposable income falls to $500 a month, the ratio would be 1.00 and the likelihood of defaulting on the payment would be 86 percent.

You might think that repossession would be an attractive option for a lender faced with a borrower who is not making payments. The collateral—in this case, the house—could be sold to pay off the loan. Given that most real estate goes up in value over time, a house that sold for $100,000 five years ago might have a much greater market value now, depending on its location. In fact, in some regions, values have recently doubled and tripled in just a few years. So, shouldn't the lender just repossess the house and sell it to another buyer to cover the remaining loan principal—and, depending on market price, perhaps make a profit?

There are some problems with that logic. The first is that some real estate markets in Maryland, Pennsylvania, and Delaware have softened recently; in fact, some real estate values have actually decreased.

Second, lenders are in the loan-making business; they are not in the real estate business. When a lender forecloses on a loan and repossesses a house, the upkeep of that house is now its problem. Maintaining an empty house is expensive, especially in the Mid-Atlantic states, where an empty house must be heated—not to mention that it must kept in good repair and the grass must be cut. Also, the house must be monitored so it is not vandalized, used by the homeless, or otherwise invaded. If the house is rented out before it is sold, the lender must collect the rent. And the lender must assume landlord obligations—fixing the plumbing, making repairs, etc. Lastly, the lender is responsible for paying real estate and other taxes.

Third, repossession is an expensive legal process. Then later reselling is also an expensive process.

Fourth, lenders who aggressively repossess collateral get a bad reputation in the community and can have trouble getting new lending business.

The bottom line is this: Teaser's management wants to make as much interest income as it can on its loans, without getting involved in significant numbers of repossessions. Thus, the company will have to do a case-by-case analysis of its loans to determine which loans require repossession and which do not. Those that do not require repossession will have their rates adjusted upward, with the amount of the increase determined, again, on a case-by-case basis.

The detailed case-by-case analyses will be done later in this chapter. First, management must gain an overall sense of the company's situation—its loans, the corresponding rates of the loans, and the profiles of the borrowers. In other words, management needs to know the following:

- The current value of the company's low-rate loans
- The value of refinanced loans at some key amounts when default is likely
- The identity of loan representatives making risky loans. If so, such loans might be refinanced at a percentage different from others.

THE COMPANY'S DATABASE

To continue this case, you must have the file LOWLOANS.accdb, which you can find in your Data files in the Case 11 folder. That file has tables of data on the bank's 1 percent teaser loans that are now reaching the 5-year mark. The database tables are discussed next. Then you will be told what is required in your analysis.

The CUSTOMERS Table

The CUSTOMERS table has data about the loan company's customers. The nature of the Cust Num, Last Name, and Initial fields should be clear. The Disposable field is the customer's current disposable income—the amount the customer has available to spend. The Property Num field shows a number corresponding to the property the customer owns, and the State field shows in what state the property is located. In addition, each customer has a loan representative, whose name is shown in the Rep field. One hundred borrowers have 1 percent loans nearing the 5-year mark. The first few records in the table are shown in Figure 11-2.

Customers

Cust Num	Last Name	Initial	Disposable	Property Num	State	Rep
C0001	Abdur-Rahim	Y	$893.00	P8000	PA	SUE
C0005	Abreu	N	$1,146.00	P8001	PA	SUE
C0006	Adams	S	$873.00	P8003	DE	JACK
C0010	Aebischer	K	$1,035.00	P8005	PA	SUE
C0011	Afanasenkov	G	$876.00	P8008	DE	JACK
C0012	Afinogenov	I	$1,044.00	P8011	MD	PAT

FIGURE 11-2 CUSTOMERS table

Data about the properties that the borrowers own are shown in the PROPERTIES table. The appraised value of the property when the mortgage was obtained five years ago is in the Appraisedfield. The value of the property as it nears the five-year mark is shown in the Current Val field. Assume each customer has one loan on one property. The first few records in the table are shown in Figure 11-3.

Properties

Property Num	Appraised	Current Val
P8000	$144,000.00	$140,500.00
P8001	$119,000.00	$138,500.00
P8003	$88,000.00	$84,500.00
P8005	$148,000.00	$148,000.00
P8008	$140,000.00	$136,000.00
P8011	$104,000.00	$113,500.00

FIGURE 11-3 PROPERTIES table

Data about the loans made are in the LOANS table. Loan Num is the primary key field. Each loan was for 20 years, comprising 240 monthly payments. Principal is the initial amount of the home loan. The monthly payment at 1 percent is shown in the Payment field. One hundred loans are nearing the five-year mark. The first few records in the table are shown in Figure 11-4.

Loans

Loan Num	Cust Num	Property Num	Principal	Payment
L2500	C0001	P8000	$142,560.00	$655.63
L2502	C0005	P8001	$117,810.00	$541.80
L2504	C0006	P8003	$87,120.00	$400.66
L2507	C0010	P8005	$146,520.00	$673.84
L2511	C0011	P8008	$138,600.00	$637.41
L2514	C0012	P8011	$102,960.00	$473.51
L2516	C0014	P8014	$134,640.00	$619.20

FIGURE 11-4 LOANS table

ASSIGNMENT 1: USING ACCESS AND EXCEL FOR DECISION SUPPORT

You will analyze the low-rate loans by working through the following sequence of steps:

1. Make a data gathering query in Access. The query output will be imported into Excel for analysis.
2. Open a blank spreadsheet called LOWLOANS.xls. Import the query output into the file's **Sheet1**.
3. Augment the worksheet to add an input zone.
4. Compute added loan data in the worksheet.
5. Add the VLOOKUP table to the worksheet.
6. Copy data to a second sheet for data table analysis.
7. Calculate default likelihoods for loans.
8. Make a table for data analysis.
9. Compute data for the portfolio.

Each of those nine steps is described in detail in the following section.

Steps to Analyze Low-Rate Loans

Step 1: Make a data gathering query in Access.

Make a query that outputs the customer number, last name, disposable income, loan representative name, state, property number, appraised value, current value, loan number, initial loan principal, and monthly payment. Note that the query should have as many output rows as there are loans. Save the query as DataGather. Close the database file.

Step 2: Import the query output into worksheet Sheet1.

In Excel, open and save a new file called LOWLOANS.xls. Import the DataGather query output into **Sheet1** by following this procedure:

1. Click the **Data** tab.
2. In the Get External Data group, select From Access. In the Select Data Source dialog box, identify LOWLOANS.accdb as the source file.
3. Select the DataGather query from the Select Table dialog box and click **OK**.
4. The data will be brought into the worksheet as a "Table." To convert the table to a data range, click a cell in the table. In Table Tools–Tools Group, select Convert to Range.
5. The cells may retain a formatted look, and you may want to change the formatting to a more standard black and white appearance. To do that, click a blank cell outside the range; then select the format painter (Home tab, Clipboard group, Format Painter) to capture the format of the empty cell. Highlight the entire data range. When you release the cursor, the appearance will change to the standard look.

Step 3: Augment the worksheet for an input zone.

Insert a number of rows (Home tab, Cells group, Insert, Insert Rows). Add an input zone for the initial loan rate (1 percent) and a new loan rate (here, 6 percent is entered to give Excel something to work with). When you are finished, the worksheet should look like the one shown in Figure 11-5.

	A	B	C	D	E	F	G	H	I	J	K
1											
2	INITIAL LOAN RATE	1%									
3											
4	NEW LOAN RATE	6%									
5											
6										INITIAL	PAYMENT
7	CUST NUM	LAST NAME	DISPOSABLE	REP	STATE	PROPERTY NUM	APPRAISED	CURRENT VAL	LOAN NUM	PRINCIPAL	AT 1%
8	C0001	Abdur-Rahim	$893	SUE	PA	P8000	$144,000	$140,500	L2500	$142,560	$655.63
9	C0005	Abreu	$1,146	SUE	PA	P8001	$119,000	$138,500	L2502	$117,810	$541.80
10	C0006	Adams	$873	JACK	DE	P8003	$88,000	$84,500	L2504	$87,120	$400.66
11	C0010	Aebischer	$1,035	SUE	PA	P8005	$148,000	$148,000	L2507	$146,520	$673.84
12	C0011	Afanasenkov	$876	JACK	DE	P8008	$140,000	$136,000	L2511	$138,600	$637.41

FIGURE 11-5   Augmented worksheet

Note that after the importing operation the Excel column headings will correspond to the database table's field names. In Figure 11-5, the column headings have been put in capital letters and bolded for readability. Bolding is done with the Home tab, Font group, B icon. Also, the field name "Principal" has been changed to the column heading "Initial Principal" and similarly "Payment" has been changed to "PAYMENT at 1%" for greater clarity in the discussions that follow. Lastly, currency value formatting has been set for the spreadsheet using the Home tab, Number group, Currency format, Dollar sign drop-down menu, More Accounting Formats choice, Currency format, set decimals as appropriate.

You should enter a border for the input cells by using Home tab, Font group, Border drop-down menu, Outside Border choice. Then format the input cells to hold numbers as a Percentage (Home tab, Number group, Number Format drop-down menu). Thus, when the number .06, for example, is entered, 6% would appear in the cell, as shown in Figure 11-5.

Step 4: Compute added loan data in the worksheet.

For each loan, you need to compute the following:

- The amount of principal paid off after the first five years of payments. Use the CUMPRINC() formula to do that. *CUMPRINC* stands for "cumulative principal." (You may need to use Office Button, Excel Options, Add-Ins—Analysis Toolpak to get Excel to add this financial function to its repertoire.)
- The amount of principal remaining on the loan. This is the initial principal, less the cumulative principal paid off during the first five years.
- New monthly payment. Use PMT() and the new interest rate input to compute the new monthly payment. The new loan is a 15-year loan. Assume payments are made monthly at the end of each month.
- The ratio of the new payment to disposable income. This is the new payment divided by a customer's disposable income.

Assume 6 percent is used as the new loan rate. The added data values would look like that shown in Figure 11-6. (Only the top few rows are shown.)

	K	L	M	N	O
1					
2					
3					NEW
4					PAYMENT
5		PRINCIPAL	REM-		TO DIS-
6	PAYMENT	PAID FIVE	MAINING	NEW	POSABLE
7	AT 1%	YEARS	PRINCIPAL	PAYMENT	RATIO
8	$655.63	-$33,014.25	$109,546	$924.41	1.035
9	$541.80	-$27,282.61	$90,527	$763.92	0.667
10	$400.66	-$20,175.37	$66,945	$564.92	0.647
11	$673.84	-$33,931.31	$112,589	$950.09	0.918

FIGURE 11-6   Added loan data values

Now you will learn how to interpret this added data, using the first loan listed in Figure 11-6 as an example. Note that the first loan corresponds to number L2500 in the LOANS table shown in Figure 11-4 and was issued for a principal of $142,560. In five years, $33,014 of the principal was paid off. A new loan would be for $142,560 less $33,014, or $109,546 in remaining principal. For the new loan's 15-year period, the monthly payment would be $924.41. Figure 11-2 shows that this customer's monthly disposable income is $893. Thus, the ratio of the new payment to the customer's disposable income is $924/$893 = 1.035.

Step 5: Add the VLOOKUP table to the worksheet.

Enter a range of data to the right of the new payment to disposable income ratios. Use the ratios and default likelihoods shown in Figure 11-1. The first few entries are shown in Figure 11-7.

O	P	Q	R
NEW			
PAYMENT			
TO DIS-			
POSABLE			
RATIO			
1.035		RATIO	LIKELIHOOD
0.667		0.20	0.01
0.647		0.25	0.02
0.918		0.30	0.03

FIGURE 11-7   Default likelihood table values

In the next worksheet, you will use this range in a VLOOKUP calculation.

Step 6: Copy the data to a second sheet, Sheet2, for data table analysis.

The remainder of the analysis will be done using a copy of the data, which preserves Sheet1 as a data source. In case some error is made, your backup will be the data in Sheet1.

Copy your data to a second worksheet. Highlight the data, except for the Ratio and Likelihood columns. In the Home tab, Clipboard group, select Copy. Then select Sheet2. Position the cursor in cell A1. Select Paste.

Step 7: Calculate default likelihoods for loans.

Use the VLOOKUP data in Sheet1 to compute default likelihoods for each loan. (To refer to a value in Sheet1, use exclamation point syntax. For example, suppose you have a sheet named X and a sheet named Y. In cell B4 of sheet X, you have an input value. You want to reference that value in cell D77 of sheet Y. The formula in D77 would refer to the input value as =SheetX!B4.)

When you are finished, the top part of your data in Sheet2 should look like that shown in Figure 11-8.

	O	P
1		
2		
3	NEW	
4	PAYMENT	
5	TO DIS-	
6	POSABLE	DEFAULT
7	RATIO	LIKELIHOOD
8	1.035	0.86
9	0.667	0.23
10	0.647	0.18
11	0.918	0.63

FIGURE 11-8 Default likelihood data for loans

Now you will learn how to interpret this added data, using the first loan listed in Figure 11-8 as an example. Note that, as in the earlier example, this first loan corresponds to loan number L2500 in the LOANS table shown in Figure 11-4. As you discovered, the new payment to disposable income ratio for this loan is 1.035. Figure 11-8 shows that 86 percent of borrowers with a loan having this ratio would fail to make their monthly payments. That suggests that the borrower of loan L2500 probably would fail to make the monthly payments if the original 20-year loan at 1 percent was reissued as a 15-year loan at 6 percent.

Step 8: Make a table for data analysis.

Follow this procedure to turn the data range into a data table:

1. In Sheet2, highlight the data area, including the header line (in the example, row 7).
2. Select the Insert tab (not the Data tab). In the Tables group, select Table. Click **OK**.
3. You need a bottom totals row in the table. To insert that row, in the Design tab, click the **Total Row** box in the Design group.

Step 9: Compute data for the portfolio.

Now you can use the table to gather the data that management wants:

1. Data about the overall portfolio at both the 1 percent rate and at the 6 percent rate
2. Data about the riskiness of loans made by each loan representative

You might be able to generate some of the needed data using built-in functions, but you are required to generate data using Excel's table.

To be specific, here is the data you need to gather:

- Number of loans and the total initial principal value of loans at 1 percent
- Number of loans and total initial value of loans for each loan representative at 1 percent
- Number and total principal of 15-year loans whose default likelihood is 50 percent or more (assume a 6 percent new loan rate)
- Number and total principal of 15-year loans at 6 percent whose default likelihood is 50 percent or more, sorted by each loan representative

Management does not know what to expect from this analysis. The analysis may show that only a few loans are expected to default, in which case the results would turn out as shown in Figure 11-9. (Values are illustrative.)

	AMOUNT
LOANS at 1%	
NUMBER	100
$ PRINCIPAL	$12,000,000
DEFAULTED 6% LOANS (>50% likelihood)	
NUMBER	10
$ PRINCIPAL	$1,000,000

FIGURE 11-9 Illustrative data

The analysis in Figure 11-9 assumes that 100 initial loans were offered at 1 percent, with a total principal of $12 million. The results show that when the interest rate increases to 6 percent, only 10 loans (valued at a total of $1 million) are 50 percent or more likely to be in default. That means, of course, that 90 loans would have less than a 50 percent likelihood of default. In this case (i.e., if only 10 out of 100 loans were likely to go into default), management might conclude that it would be worth refinancing all loans at 6 percent.

But the results might turn out as shown in Figure 11-10.

	AMOUNT
LOANS at 1%	
NUMBER	100
$ PRINCIPAL	$12,000,000
DEFAULTED 6% LOANS (>50% likelihood)	
NUMBER	45
$ PRINCIPAL	$6,000,000

FIGURE 11-10 Illustrative data

The analysis in Figure 11-10 suggests that 45 loans (valued at a total of $6 million) would be 50 percent or more likely to be in default. In that case, with so many loans likely to go into default, management might conclude that it would not be worth refinancing all loans at 6 percent. Instead, management might opt for a lower interest rate increase or it might opt to refinance only those loans that seem unlikely to default.

The other sort of information management wants to know is whether any risk is associated with individual loan representatives, as opposed to the portfolio as a whole. The results of that type of analysis might turn out as shown in Figure 11-11.

	SUE	PAT	JACK	TOTAL
LOANS at 1%				
NUMBER	30	30	40	100
$ PRINCIPAL	$2.0M	$3.0M	$7.0M	$12.0M
DEFAULTED 6% LOANS (>50% likelihood)				
NUMBER	7	8	30	45
$ PRINCIPAL	$.4M	$.6M	$5.0M	$6.0M

FIGURE 11-11   Illustrative data

In Figure 11-11, it appears again that $6 million of loans will go into default. But most of those are Jack's loans. Management might be inclined to refinance Sue's and Pat's loans, which look relatively less risky, and to deal separately with Jack's loans (and with Jack).

As you work with the data table, you can manually copy and paste interesting results as notes in Sheet2 for later reference when you write your memo.

ASSIGNMENT 2: USING THE SPREADSHEET FOR DECISION SUPPORT

You are now in a position to summarize the results. You will write a memo in Word, summarizing the results. When you are done with the spreadsheet, follow these steps:

1. Save the file one last time (Office Button—Save).
2. Click **Office Button—Close** and **Exit**. Only then should you remove your disk from drive A: or remove your USB flash drive.

In preparing your memo, observe the following requirements:

- Include a proper heading (TO/FROM/DATE/SUBJECT). You may use a Word memo template (Click **Office Button**, click **New**, click the **Memo** template in the Templates section, choose Contemporary Memo, then click **OK**).
- Briefly state the purpose of the analysis, but do not provide much background. Assume the reader expects the memo and knows you are working on the analysis.
- State the results of your analysis. How risky is the overall portfolio in terms of the number of loans with a default likelihood of 50 percent or more? Does it look as though one (or more) of the reps is making a disproportionate number of risky loans? What is your overall recommendation to management?

Support your analysis graphically. Insert a table in your memo that shows appropriate totals and percentages. Use the following procedure:

1. Click the cursor where you want the table to appear in the document.
2. In the Insert group, select the Insert drop-down menu.
3. Select the Insert Table.
4. Choose the number of rows and columns.
5. Click **OK**.

Your table should look like that shown in Figure 11-12.

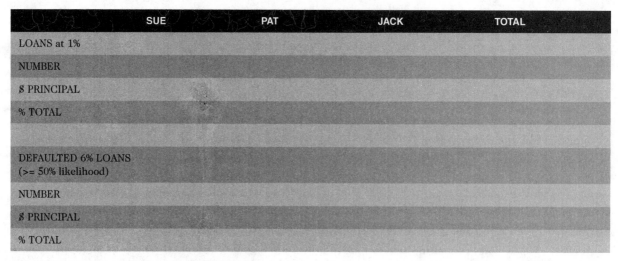

	SUE	PAT	JACK	TOTAL
LOANS at 1%				
NUMBER				
$ PRINCIPAL				
% TOTAL				
DEFAULTED 6% LOANS (>= 50% likelihood)				
NUMBER				
$ PRINCIPAL				
% TOTAL				

FIGURE 11-12 Form for table in memo

ASSIGNMENT 3: GIVING AN ORAL PRESENTATION

Assume Teaser's senior loan representative is impressed by your analysis and she thinks the rest of the company's loan representatives need to know more about it. She asks you to give a presentation explaining your program and your method. Prepare to explain your work and your recommendation to the group in ten minutes or less. Use visual aids or handouts that you think are appropriate. (Tutorial F has information on how to prepare and give an oral presentation.)

DELIVERABLES

Assemble the following deliverables for your instructor:

1. Printout of your memo
2. Disk that contains your Word, Excel, and Access files. Do not provide a CD, as it would be read-only.

If there are other .xlsx files or .accdb files on your disk, write your instructor a note identifying the names of the files that are pertinent to this case.

PART **6**

ADVANCED EXCEL SKILLS
USING EXCEL

GUIDANCE FOR EXCEL CASES

Some of the cases in this casebook require the use of some advanced techniques in Excel. Those techniques are explained here rather than in the cases themselves:

- VLOOKUP Function
- PMT Function
- CUMPRINC Function
- Data Tables

VLOOKUP FUNCTION

The logic and use of IF statements are discussed in Tutorial C. IF statements let you program the logic of a decision. Sometimes the number of possible choices associated with the decision is too great for the IF statement to handle easily. VLOOKUP is made for such situations. Here is an example of how VLOOKUP works.

Suppose a bank charges interest on loans at a rate based on the amount of the loan. The rate is computed by spreadsheet, as shown in Figure E-1.

	A	B	C
1	**VLOOKUP EXAMPLE**		
2			
3	INPUT: LOAN AMOUNT		
4			
5	RESULT: INTEREST RATE		#N/A
6			
7	LOOKUP TABLE		
8		**AMOUNT**	**RATE**
9		100	0.040
10		200	0.041
11		300	0.042
12		400	0.043
13		500	0.044
14		600	0.045
15		700	0.046
16		800	0.047
17		900	0.048

FIGURE E-1 VLOOKUP function example

As you can see in Figure E-1, the loan amounts and their corresponding rates are shown in a "lookup table" in the range B9..C17. In this spreadsheet, the user enters the loan amount and the rate is computed by the VLOOKUP formula in cell C5. (Note that the #N/A currently showing in cell C5 is not an error message—it appears because no loan amount has been entered yet.) Now assume the user enters 700 in the input cell. The function looks up the correct rate and puts it in cell C5, as shown in Figure E-2.

	A	B	C
1	**VLOOKUP EXAMPLE**		
2			
3	INPUT: LOAN AMOUNT		700
4			
5	RESULT: INTEREST RATE		0.046
6			
7	LOOKUP TABLE		
8		**AMOUNT**	**RATE**
9		100	0.040
10		200	0.041
11		300	0.042
12		400	0.043
13		500	0.044
14		600	0.045
15		700	0.046
16		800	0.047
17		900	0.048

FIGURE E-2   VLOOKUP function example, continued

The VLOOKUP function in cell C5 did just what a person would do with this table. The function went down the Amount column until it found the loan amount (here, 700). When it found that amount, it looked to the right and got the rate. The VLOOKUP function then put the rate it found (here, 0.046) in cell C5.

If the user had entered 800 in cell C3, the rate would be different, as shown in Figure E-3.

	A	B	C
1	**VLOOKUP EXAMPLE**		
2			
3	INPUT: LOAN AMOUNT		800
4			
5	RESULT: INTEREST RATE		0.047
6			
7	LOOKUP TABLE		
8		**AMOUNT**	**RATE**
9		100	0.040
10		200	0.041
11		300	0.042
12		400	0.043
13		500	0.044
14		600	0.045
15		700	0.046
16		800	0.047
17		900	0.048

FIGURE E-3   VLOOKUP function example, continued

The syntax of the VLOOKUP function is as follows:

VLOOKUP(Search value, lookup table range, number of column holding answer, T/F)

In the current example, the search value is in the input cell, C3. The table range is B9..C17. (Note that the column headers are not included in this range.) Column 2 in the range holds the answer value, the rate. True or False is an optional flag. If left out, True is assumed. True means that an exact match in the lookup table is not needed. False means that an exact match is needed. The code for this example's VLOOKUP function is shown in Figure E-4.

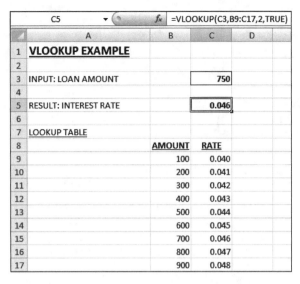

	A	B	C	D
			C5	=VLOOKUP(C3,B9:C17,2,TRUE)
1	**VLOOKUP EXAMPLE**			
2				
3	INPUT: LOAN AMOUNT		750	
4				
5	RESULT: INTEREST RATE		0.046	
6				
7	LOOKUP TABLE			
8		AMOUNT	RATE	
9		100	0.040	
10		200	0.041	
11		300	0.042	
12		400	0.043	
13		500	0.044	
14		600	0.045	
15		700	0.046	
16		800	0.047	
17		900	0.048	

FIGURE E-4   VLOOKUP function example code

When True is set, the match is made with the largest value in column 1 that is less than or equal to the search value (here, the input value in C3). Thus, if 750 were entered, a match would be made with 700, as shown in Figure E-4. If an exact match were required, the value of the fourth element of the VLOOKUP formula would be set to False. Now if an exact match cannot be made, #N/A is output, as shown in Figure E-5.

	A	B	C	D	E
			C5	=VLOOKUP(C3,B9:C17,2,FALSE)	
1	**VLOOKUP EXAMPLE**				
2					
3	INPUT: LOAN AMOUNT		750		
4					
5	RESULT: INTEREST RATE		#N/A		
6					
7	LOOKUP TABLE				
8		AMOUNT	RATE		
9		100	0.040		
10		200	0.041		
11		300	0.042		
12		400	0.043		
13		500	0.044		
14		600	0.045		
15		700	0.046		
16		800	0.047		
17		900	0.048		

FIGURE E-5   VLOOKUP function example, not exact match

PMT FUNCTION

The PMT function computes the payment for a loan based on constant payments and a constant interest rate. Assume a friend lends you $10,000 at 10 percent interest to be paid back in three equal yearly payments made at the end of each of the loan's three years. How much should each of the three payments be? The situation is shown in Figure E-6.

	A	B	C	D
1		ANNUAL		
2	LOAN	INTEREST	NUMBER	
3	AMOUNT	RATE	OF YEARS	PAYMENT
4	$10,000	10.00%	3	

FIGURE E-6 PMT function example

You want to compute the yearly payment and show it in cell D4. The PMT function is designed specifically to handle this kind of problem. The syntax of the function is as follows:

PMT(Interest Rate, Number of Periods, Present Value, Future Value, Type)

Interest Rate is the rate in each payment period. Number of Periods is the same as the number of payments—there is one payment per period. Present Value is the value in today's dollars of all of the payments; it is the same as the loan principal, assuming the loan is paid off by the payments, as is generally the case. Future Value is the balance owed when the last payment is made; this is zero if the loan is to be paid off by the payments. Type indicates whether payments are to be made at the beginning or end of the period. A value of 0 indicates the end of the period (the typical situation), and a value of 1 indicates the beginning.

In the current example, three equal annual year-end payments at 10 percent are required to pay off the $10,000 loan. The PMT code and the resulting payment calculation are shown in Figure E-7.

D4			f_x	=PMT(B4,C4,A4,0,0)	
	A	B	C	D	E
1		ANNUAL			
2	LOAN	INTEREST	NUMBER		
3	AMOUNT	RATE	OF YEARS	PAYMENT	
4	$10,000	10.00%	3	-$4,021.15	

FIGURE E-7 PMT function example

Note in Figure E-7 that instead of raw values being entered for interest rate, principal, and number of payments in the PMT formula, the cell addresses are used. Also note that Excel shows the payment as a negative number. The amount could be made positive by multiplying the PMT formula by -1.

What if the payments had been at the end of each month in the three years? That would have changed both the number of payments and the interest rate in each period. The number of payments would have been 12 times the number of years (here, 12 * 3 = 36). The interest rate for each month would have been the annual interest rate divided by 12. The PMT code and the resulting payment calculation for a monthly payment scenario are shown in Figure E-8.

D4			f_x	=PMT(B4/12,C4*12,A4,0,0)	
	A	B	C	D	E
1		ANNUAL			
2	LOAN	INTEREST	NUMBER		
3	AMOUNT	RATE	OF YEARS	PAYMENT	
4	$10,000	10.00%	3	-$322.67	

FIGURE E-8 PMT function example, monthly basis

CUMPRINC FUNCTION

The name of the CUMPRINC function stands for "cumulative principal." The word *cumulative* denotes an accumulation over some period of time—in other words, a sum. The CUMPRINC function computes the amount of loan principal paid off between two periods.

To appreciate how the CUMPRINC function can be applied, you first need to understand how loan principal is reduced as loan payments are made. Return to the example involving a three-year loan for $10,000. Payments each year were $4,021. The beginning loan balance was $10,000. How much interest was included in the first year's payment? Why, $10,000 times 10 percent, or $1,000, of course. Then how much of that first-year payment was devoted to reducing the amount of principal owed? ($4,021 − $1,000 = $3,021) Is the amount of interest and the amount of principal paid the same for each year's payment? No, because the principal owed constantly decreases. For the example loan, the interest and principal of each payment is shown in Figure E-9.

Year	Beginning Principal Owed	Payment	Interest Paid at 10%	Principal Paid	Principal Owed, End of Year
1	10,000	4,021	1,000	3,021	6,979
2	6,979	4,021	698	3,323	3,656
3	3,656	4,021	365	3,656	0

FIGURE E-9 Payment of example loan in three years

Notice that each year's $4,021 payment has less interest in it than the year before and more principal in it than the year before.

How much principal was paid off in years 1 and 2? ($3,021 + $3,323 = $6,344) How much principal was paid off in years 2 and 3? ($3,323 + $3,656 = $6,979) The CUMPRINC function is specifically designed to answer those sorts of questions. The syntax of the CUMPRINC function is as follows:

CUMPRINC(Interest Rate, Number of Periods, Present Value, Start Period, End Period, Type)

Interest Rate, Number of Periods, and Present Value have the same meaning as in the PMT function. Start Period and End Period are the loan payment periods in question. With respect to the previous question regarding how much principal would be paid off after two years, the Start and End Periods would be 1 and 2. Figure E-10 shows the three-year $10,000 loan example with a column added for principal paid after two years.

	A	B	C	D	E
1		ANNUAL			PRIN PAID
2	LOAN	INTEREST	NUMBER		AFTER
3	AMOUNT	RATE	OF YEARS	PAYMENT	TWO YEARS
4	$10,000	10.00%	3	-$4,021.15	

FIGURE E-10 CUMPRINC function example

The CUMPRINC function is used in cell E4 to compute the amount of principal paid after two years, as shown in Figure E-11.

E4		fx	=CUMPRINC(B4,C4,A4,1,2,0)		
	A	B	C	D	E
1		ANNUAL			PRIN PAID
2	LOAN	INTEREST	NUMBER		AFTER
3	AMOUNT	RATE	OF YEARS	PAYMENT	TWO YEARS
4	$10,000	10.00%	3	-$4,021.15	-$6,344.41

FIGURE E-11 CUMPRINC function example, continued

Notice that the function shows an amount as a negative number. Negative numbers aside, you can see that the amount produced by the CUMPRINC function—$6,344—is the same amount as was tabulated using Figure E-9 (pennies excepted). Figure E-12 shows the example extended to cover years 2 and 3 (the last two years).

F4				f_x	=CUMPRINC(B4,C4,A4,2,3,0)	

	A	B	C	D	E	F
1		ANNUAL			PRIN PAID	PRIN PAID
2	LOAN	INTEREST	NUMBER		AFTER	IN LAST
3	AMOUNT	RATE	OF YEARS	PAYMENT	TWO YEARS	TWO YEARS
4	$10,000	10.00%	3	-$4,021.15	-$6,344.41	-$6,978.85

FIGURE E-12 CUMPRINC function example, continued

The resulting amount is $6,979 (rounded), as shown in Figure E-9.

DATA TABLE

An Excel data table is a contiguous range of data that has been designated as a table. Once so designated, the table gains certain properties that are useful for data analysis. (*Note:* In previous versions of Excel, data tables were called *data lists*.) Suppose you have a list of runners who have run a race, as shown in Figure E-13.

	A	B	C	D	E	F
1	RUNNER#	LAST	FIRST	AGE	GENDER	TIME (MIN)
2	100	HARRIS	JANE	O	F	70
3	101	HILL	GLENN	Y	M	70
4	102	GARCIA	PEDRO	M	M	85
5	103	HILBERT	DORIS	M	F	90
6	104	DOAKS	SALLY	Y	F	94
7	105	JONES	SUE	Y	F	95
8	106	SMITH	PETE	M	M	100
9	107	DOE	JANE	O	F	100
10	108	BRADY	PETE	O	M	100
11	109	BRADY	JOE	O	M	120
12	110	HEEBER	SALLY	M	F	125
13	111	DOLTZ	HAL	O	M	130
14	112	PEEBLES	AL	Y	M	63

FIGURE E-13 Data table example

To turn that information into a data table (list), you highlight the data range, including headers, and select the Insert tab, then in the Tables group, you click **Table**. Once you do that, you get the Create Table window, as shown in Figure E-14.

A1				f_x	RUNNER#	

	A	B	C	D	E	F
1	RUNNER#	LAST	FIRST	AGE	GENDER	TIME (MIN)
2	100	HARRIS	JANE	O	F	70
3	101	HILL	GLENN	Y	M	70
4	102	GARCI				85
5	103	HILBER				90
6	104	DOAKS				94
7	105	JONES				95
8	106	SMITH				100
9	107	DOE				100
10	108	BRADY				100
11	109	BRADY	JOE	O	M	120
12	110	HEEBER	SALLY	M	F	125
13	111	DOLTZ	HAL	O	M	130
14	112	PEEBLES	AL	Y	M	63

Create Table

Where is the data for your table?

=$A$1:$F$14

☑ My table has headers

OK   Cancel

FIGURE E-14 Create Table window

You then click **OK**. The data range will now show as a table. In the Design tab, you select Total Row to add a totals row to the data table. You also can select Table Styles – Light to get rid of the contrasting color in the table's rows. Figure E-15 shows the results of doing those steps.

	A	B	C	D	E	F
1	RUNNER#	LAST	FIRST	AGE	GENDER	TIME (MIN)
2	100	HARRIS	JANE	O	F	70
3	101	HILL	GLENN	Y	M	70
4	102	GARCIA	PEDRO	M	M	85
5	103	HILBERT	DORIS	M	F	90
6	104	DOAKS	SALLY	Y	F	94
7	105	JONES	SUE	Y	F	95
8	106	SMITH	PETE	M	M	100
9	107	DOE	JANE	O	F	100
10	108	BRADY	PETE	O	M	100
11	109	BRADY	JOE	O	M	120
12	110	HEEBER	SALLY	M	F	125
13	111	DOLTZ	HAL	O	M	130
14	112	PEEBLES	AL	Y	M	63
15	Total					1242

FIGURE E-15 Data table example

The headers have acquired drop-down menu tabs, as you can see in Figure E-15.

You can sort the data table records by any field. Perhaps you want to sort by Times. You would click the drop-down menu in that header and select Sort – Smallest to Largest. You would get the results shown in Figure E-16.

	A	B	C	D	E	F
1	RUNNER	LAST	FIRST	AGE	GENDE	TIME (MIN)
2	112	PEEBLES	AL	Y	M	63
3	100	HARRIS	JANE	O	F	70
4	101	HILL	GLENN	Y	M	70
5	102	GARCIA	PEDRO	M	M	85
6	103	HILBERT	DORIS	M	F	90
7	104	DOAKS	SALLY	Y	F	94
8	105	JONES	SUE	Y	F	95
9	106	SMITH	PETE	M	M	100
10	107	DOE	JANE	O	F	100
11	108	BRADY	PETE	O	M	100
12	109	BRADY	JOE	O	M	120
13	110	HEEBER	SALLY	M	F	125
14	111	DOLTZ	HAL	O	M	130

FIGURE E-16 Sorting list by drop-down menu

By doing that, you see that Peebles had the best time and Doltz had the worst time. You also can sort Largest to Smallest.

In addition, you can sort by more than one criteria. Assume you want to sort first by Gender and then by Time (within Gender). You first Sort Smallest to Largest in Gender. Then you again activate the Gender drop-down tab and select Sort By Color – Custom Sort. In the Sort window that appears, you click Add Level and choose Time as the next criteria. That is shown in Figure E-17.

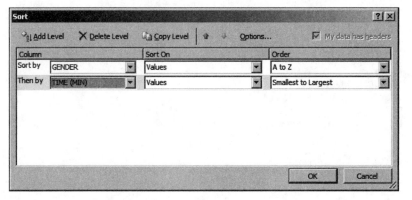

FIGURE E-17 Sorting on multiple criteria

You then click the OK button and get the results shown in Figure E-18.

	A	B	C	D	E	F
1	RUNNER	LAST	FIRST	AGE	GENDE	TIME (MIN
2	100	HARRIS	JANE	O	F	70
3	103	HILBERT	DORIS	M	F	90
4	104	DOAKS	SALLY	Y	F	94
5	105	JONES	SUE	Y	F	95
6	107	DOE	JANE	O	F	100
7	110	HEEBER	SALLY	M	F	125
8	112	PEEBLES	AL	Y	M	63
9	101	HILL	GLENN	Y	M	70
10	102	GARCIA	PEDRO	M	M	85
11	106	SMITH	PETE	M	M	100
12	108	BRADY	PETE	O	M	100
13	109	BRADY	JOE	O	M	120
14	111	DOLTZ	HAL	O	M	130

FIGURE E-18 Sorting by Gender and Time (within Gender)

By doing that, you can see that Harris had the best female time and that Peebles had the best male time.

Perhaps you want to see the Top *N* in some criteria, for example the top five times. You select the Time criteria's drop-down menu and select Number Filters. You get another menu, in which you select Top 10. You get the Top 10 AutoFilter window, as shown in Figure E-19.

FIGURE E-19 Top 10 AutoFilter

This window lets you specify the number of values you want. In Figure E-19, five values were entered. Clicking OK, you would get the results shown in Figure E-20.

	A	B	C	D	E	F
1	RUNNER ▾	LAST ▾	FIRST ▾	AGE ▾	GENDE ▾	TIME (MIN ▾
6	107 DOE		JANE	O	F	100
7	110 HEEBER		SALLY	M	F	125
11	106 SMITH		PETE	M	M	100
12	108 BRADY		PETE	O	M	100
13	109 BRADY		JOE	O	M	120
14	111 DOLTZ		HAL	O	M	130
15	Total					675

FIGURE E-20   Top 5 Times

In the output, there are more than five data records. That is because there are ties at 100 minutes. If you wanted to see all of the records again, you would click the Time drop-down menu and select Clear Filter. You then would have the full table of data back, as shown in Figure E-21.

	A	B	C	D	E	F
1	RUNNER ▾	LAST ▾	FIRST ▾	AGE ▾	GENDE ▾	TIME (MIN ▾
2	100 HARRIS		JANE	O	F	70
3	103 HILBERT		DORIS	M	F	90
4	104 DOAKS		SALLY	Y	F	94
5	105 JONES		SUE	Y	F	95
6	107 DOE		JANE	O	F	100
7	110 HEEBER		SALLY	M	F	125
8	112 PEEBLES		AL	Y	M	63
9	101 HILL		GLENN	Y	M	70
10	102 GARCIA		PEDRO	M	M	85
11	106 SMITH		PETE	M	M	100
12	108 BRADY		PETE	O	M	100
13	109 BRADY		JOE	O	M	120
14	111 DOLTZ		HAL	O	M	130
15	Total					1242

FIGURE E-21   Restoring all data to screen

Each of the cells in the Total row has a drop-down menu. The menu choices are statistical operations that you can do on the totals—take a Sum; take an Average, Min, Max; Count the number of records; and others. Assume the Time drop-down menu was selected as shown in Figure E-22.

	A	B	C	D	E	F
1	RUNNER	LAST	FIRST	AGE	GENDE	TIME (MIN
2	100	HARRIS	JANE	O	F	70
3	103	HILBERT	DORIS	M	F	90
4	104	DOAKS	SALLY	Y	F	94
5	105	JONES	SUE	Y	F	95
6	107	DOE	JANE	O	F	100
7	110	HEEBER	SALLY	M	F	125
8	112	PEEBLES	AL	Y	M	63
9	101	HILL	GLENN	Y	M	70
10	102	GARCIA	PEDRO	M	M	85
11	106	SMITH	PETE	M	M	100
12	108	BRADY	PETE	O	M	100
13	109	BRADY	JOE	O	M	120
14	111	DOLTZ	HAL	O	M	130
15	Total					1242
16						None
17						Average
18						Count
19						Count Numbers
20						Max
21						Min
22						Sum
						StdDev
						Var
						More Functions…

FIGURE E-22 Selecting Time drop-down menu in Total row

Using the Average function, you find that the average time for all runners was 95.5 minutes, as shown in Figure E-23.

	A	B	C	D	E	F
1	RUNNER	LAST	FIRST	AGE	GENDE	TIME (MIN
2	100	HARRIS	JANE	O	F	70
3	103	HILBERT	DORIS	M	F	90
4	104	DOAKS	SALLY	Y	F	94
5	105	JONES	SUE	Y	F	95
6	107	DOE	JANE	O	F	100
7	110	HEEBER	SALLY	M	F	125
8	112	PEEBLES	AL	Y	M	63
9	101	HILL	GLENN	Y	M	70
10	102	GARCIA	PEDRO	M	M	85
11	106	SMITH	PETE	M	M	100
12	108	BRADY	PETE	O	M	100
13	109	BRADY	JOE	O	M	120
14	111	DOLTZ	HAL	O	M	130
15	Total					95.53846154

FIGURE E-23 Average running time shown in Total row

PART 7

PRESENTATION SKILLS

TUTORIAL **F**

GIVING AN ORAL PRESENTATION

Giving an oral presentation provides you the opportunity to practice the presentation skills you'll need in the workplace. The presentations you create for the cases in this textbook will be similar to real-world presentations. You'll present objective, technical results to an organization's stakeholders, and you'll support your presentation with visual aids commonly used in the business world. During your presentation, your instructor might assign your classmates to role-play an audience of business managers, bankers, or employees and ask them give you feedback on your presentation.

Follow these four steps to create an effective presentation:

1. Plan your presentation.
2. Draft your presentation.
3. Create graphics and other visual aids.
4. Practice your delivery.

You'll start at the beginning and look at the steps involved in planning your presentation.

PLANNING YOUR PRESENTATION

When planning an oral presentation, you need to be aware of your time limits, establish your purpose, analyze your audience, and gather information. This section will look at each of those elements.

Knowing Your Time Limits

You need to consider your time limits on two levels. First, consider how much time you'll have to deliver your presentation. For example, what can you expect to accomplish in ten minutes? The element of time is the driver of any presentation. It limits the breadth and depth of your talk—and the number of visual aids that you can use. Second, consider how much time you'll need for the actual process of preparing your presentation—that is, for drafting your presentation, creating graphics, and practicing your delivery.

Establishing Your Purpose

After considering your time limits, you must define your purpose: what you need and want to say and to whom you will say it. For the cases in the Access portion of this book, your purpose will be to inform and explain. For instance, a business's owners, managers, and employees may need to know how their organization's database is organized and how they could use it to fill in input forms and create reports. In contrast, for the cases in the Excel portion of the book, your purpose will be to recommend a course of action. You'll be making recommendations to business owners, managers, and bankers based on the results you obtained from inputting and running various scenarios.

Analyzing Your Audience

Once you have established the purpose of your presentation, you should analyze your audience. Ask yourself these questions: What does my audience already know about the subject? What do the audience members want to know? What do they need to know? Do they have any biases that I should consider? What level of technical detail is best suited to their level of knowledge and interest?

In some Access cases, you will make a presentation to an audience who might not be familiar with Access or with databases in general. In other cases, you might be giving a presentation to a business owner who started to work on the database but was not able to finish it. Tailor your presentation to suit your audience.

For the Excel cases, you will be interpreting results for an audience of bankers and business managers. The audience will not need to know the detailed technical aspects of how you generated your results. However, those listeners will need to know what assumptions you made prior to developing your spreadsheet, because those assumptions might have an impact on their opinion of your results.

Gathering Information

Because you will have just completed a case as you begin preparing your oral presentation, you'll have the basic information you need. For the Access cases, you should review the main points of the case and your goals. Make sure you include all of the points you think are important for the audience to understand. In addition, you might want to go beyond the requirements and explain additional ways in which the database could be used to benefit the organization now or in the future.

For the Excel cases, you can refer to the tutorials for assistance in interpreting the results from your spreadsheet analysis. For some cases, you might want to research the Internet for business trends or background information that you can use to support your presentation.

DRAFTING YOUR PRESENTATION

Now that you have completed the planning stage, you are ready to begin drafting your presentation. At this point, you might be tempted to write your presentation and then memorize it word for word. If you do, your presentation will sound unnatural because when people speak, they use a simpler vocabulary and shorter sentences than when they write. Thus, you might consider drafting your presentation by simply noting key phrases and statistics. When drafting your presentation, follow this sequence:

1. Write the main body of your presentation.
2. Write the introduction to your presentation.
3. Write the conclusion to your presentation.

Writing the Main Body

When you draft your presentation, write the body first. If you try to write the opening paragraph first, you'll spend an inordinate amount of time creating a "perfect" paragraph—only to revise it after you've written the body of your presentation.

Keeping Your Audience in Mind

To write the main body, review your purpose and your audience's profile. What are the main points you need to make? What are your audience's wants, needs, interests, and technical expertise? It's important to include some basic technical details in your presentation, but keep in mind the technical expertise of your audience.

What if your audience consists of people with different needs, interests, and levels of expertise? For example, in the Access cases, an employee might want to know how to input information into a form, but the business owner might already know how to input data and, therefore, be more interested in learning how to generate queries and reports. You'll need to acknowledge their differences in your presentation. For example, you might say, "And now, let's look at how data entry clerks can input data into the form."

Similarly, in the Excel cases, your audience will usually consist of business owners, managers, and bankers. The owners' and managers' concerns will be profitability and growth. In contrast, the bankers' main concern will be repayment of a loan. You'll need to address the interests of each group.

Using Transitions and Repetition

Because your audience can't read the text of your presentation, you'll need to use transitions to compensate. Words such as *next*, *first*, *second*, and *finally* will help your audience follow the sequence of your ideas. Words such as *however*, *in contrast*, *on the other hand*, and *similarly* will help the audience follow shifts in thought. You also can use your voice and hand gestures to convey emphasis.

Also think about how you can use body language to emphasize what you're saying. For instance, if you are stating three reasons, you can use your fingers to tick off each reason as you discuss it: one, two, three. Similarly, if you're saying that profits will be flat, you can make a level motion with your hand for emphasis.

As you draft your presentation, repeat key points to emphasize them. For example, suppose your point is that outsourcing labor will provide the greatest gains in net income. Begin by previewing that concept. State that you're going to demonstrate how outsourcing labor will yield the biggest profits. Then provide statistics that support your claim and show visual aids that graphically illustrate your point. Summarize by repeating your point: "As you can see, outsourcing labor does yield the biggest profits."

Relying on Graphics to Support Your Talk

As you write the main body, think of how you can best incorporate graphics into your presentation. Don't waste words describing what you're presenting when you can use a graphic to portray the subject quickly. For instance, instead of describing how information from a query is input into a report, show a sample, a query result, and a completed report. Figure F-1 and Figure F-2 show an Access query and the resulting report, respectively.

Customer Name	Customer City	Product Name	Quantity	Total Value
Altamar	Miami	Xlarge	15	$825.00
Café Pacific	Dallas	Medium	10	$300.00
Café Pacific	Dallas	Large	10	$420.00
Café Pacific	Dallas	Xlarge	10	$550.00
Red Lobster	Wichita	Xlarge	15	$825.00
Zee Grill	Toronto	Large	25	$1,050.00
Big Fish	Dearborn	Medium	5	$150.00
Big Fish	Dearborn	Large	5	$210.00
Big Fish	Dearborn	Xlarge	5	$275.00
The Park Hotel	Charlotte	Medium	3	$90.00
St. Elmo Steak House	Indianapolis	Large	10	$420.00
Bound'Ry Restaurant	Nashville	Xlarge	15	$825.00
Bellagio	Las Vegas	Large	5	$210.00
Osetra The Fish House	San Diego	Medium	12	$360.00

FIGURE F-1   Access query

Weekly Sales

Customer Name	Customer City	Product Name	Quantity	Total Value
Altamar	Miami			
		Xlarge	15	$825.00
Total Quantity and Value			15	$825.00
Bellagio	Las Vegas			
		Large	5	$210.00
Total Quantity and Value			5	$210.00
Big Fish	Dearbom			
		Xlarge	5	$275.00
		Large	5	$210.00
		Medium	5	$150.00
Total Quantity and Value			15	$635.00
Bound'Ry Restaurant	Nashville			
		Xlarge	15	$825.00
Total Quantity and Value			15	$825.00
Café Pacific	Dallas			
		Xlarge	10	$550.00
		Large	10	$420.00
		Medium	10	$300.00
Total Quantity and Value			30	$1,270.00
Osetra The Fish House	San Diego			

FIGURE F-2   Access report

Also consider what kinds of graphics media are available—and how well you can use them. For example, if you've never used Microsoft PowerPoint to prepare a presentation, will you have enough time to learn the software before you deliver your upcoming presentation? (If you don't know how to use PowerPoint, you might consider finding a tutorial on the Web to help you learn the basics.)

Anticipating the Unexpected

Even though you're just drafting your presentation at this stage, eventually you'll be answering questions from the audience. Being able to handle questions smoothly is the mark of a professional. The first steps to addressing audience members' questions are being able to anticipate them and preparing your answers.

You won't use all of the facts you gather for your presentation. However, as you draft your presentation, you might want to jot down some of those facts and keep them handy—just in case you need them to answer questions from the audience. For instance, during some Excel presentations, you might be asked why you are not recommending a certain course of action or why you did not mention that course of action in your report.

Writing the Introduction

After you have written the main body of your talk, you will want to develop an introduction. An introduction should be only a paragraph or two in length and should preview the main points that your presentation will cover.

For some of the Access cases, you might want to include general information about databases: what they can do, why they are used, and how they can help the company become more efficient and profitable. You won't need to say much about the business operation because the audience already works for the company.

For the Excel cases, you might want to include an introduction to the general business scenario and describe any assumptions you made when creating and running your decision support spreadsheet. Excel is used for decision support, so you should describe the choices and decision criteria you faced.

Writing the Conclusion

Every good presentation needs a good ending. Don't leave the audience hanging. Your conclusion should be brief—only a paragraph or two—and it should give your presentation a sense of closure. Use the conclusion to repeat your main points or, for the Excel cases, to recap your findings and/or recommendations.

CREATING GRAPHICS

Using visual aids is a powerful means of getting your point across and making it understandable to your audience. Visual aids come in a variety of forms, some of which are more effective than others.

Choosing Graphics Media

The media you use should depend on your situation and the media you have available. One of the key points to remember when using any media is this: *You must maintain control of the media, or you'll lose control of your audience.*

The following list highlights some of the most common media and their strengths and weaknesses:

- **Handouts:** This medium is readily available in classrooms and in businesses. It relieves the audience from taking notes. The graphics in handouts can be multicolored and of professional quality. *Negatives:* You must stop and take time to hand out individual pages. During your presentation, the audience may study and discuss your handouts rather than listen to you. Lack of media control is *the* major drawback—and it can hurt your presentation.
- **Chalkboard (or whiteboard):** This informal medium is readily available in the classroom but not in many businesses. *Negatives:* You need to turn your back to the audience when you write (thereby running the risk of losing the audience's attention), and you need to erase what you've written as you continue your presentation. With chalkboards, your handwriting must be good—that is, it should be legible even when you write quickly. In addition, attractive graphics are difficult to create.
- **Flip Chart:** This informal medium is readily available in many businesses. *Negatives:* The writing space is so small that it's effective only for a very small audience. This medium shares many of the same negatives as the chalkboard.
- **Overheads:** This medium is readily available in classrooms and in businesses. You have control over what the audience sees and when they see it. You can create professional PowerPoint presentations

on overhead transparencies. *Negatives:* If you don't use an application such as PowerPoint, handwritten overheads look amateurish. Without special equipment, professional-looking graphics are difficult to prepare.

- **Slides:** This formal medium is readily available in many businesses and can be used in large rooms. You can use 35 mm slides or the more popular electronic on-screen slides. In fact, electronic on-screen slides are usually *the* medium of choice for large organizations and are generally preferred for formal presentations. *Negatives:* You must have access to the equipment needed for slide presentations and know how to use it. It takes time to learn how to create and use computer graphics. Also, you must have some source of ambient light; otherwise, may have difficulty seeing your notes in the dark.

Creating Charts and Graphs

Technically, charts and graphs are not the same thing, although many graphs are referred to as charts. Usually charts show relationships and graphs show change. However, Excel makes no distinction and calls both entities charts.

Charts are easy to create in Excel. Unfortunately, they are so easy to create that people often use graphics that are meaningless or that inaccurately reflect the data the graphics represent. Next, you'll look at how to select the most appropriate graphics.

Creating Charts

You should use pie charts to display data that is related to a whole. For example, you might use a pie chart when showing the percentage of shoppers who bought a generic brand of toothpaste versus a major brand, as shown in Figure F-3. (Note that when creating a pie chart, Excel takes the numbers you want to graph and makes them a percentage of 100.) You would *not*, however, use a pie chart to show a company's net income over a three-year period. While Figure F-4 does show such a pie chart, that graphic is not meaningful because it is not useful to think of the period as "a whole" or the years as its "parts."

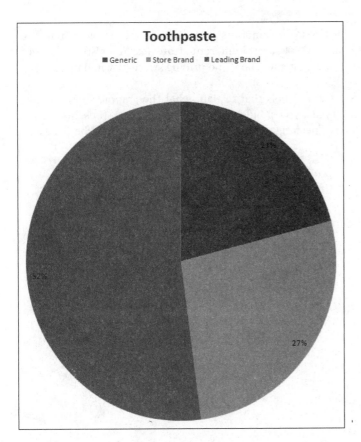

FIGURE F-3 Pie chart: appropriate use

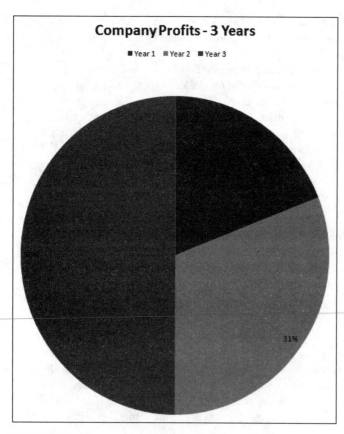

Company Profits - 3 Years

■ Year 1　■ Year 2　■ Year 3

50%

31%

FIGURE F-4　Pie chart: inappropriate use

You should use bar charts when you want to compare several amounts at one time. For example, you might want to compare the net profit that would result from each of several different strategies. You also can use a bar chart to show changes over time. For example, you might show how one pricing strategy would increase profits year after year.

When you are showing a graphic, you need to include labels that explain what the graphic shows. For instance, when you're using a graph with an x- and y-axis, you should show what each axis represents so the audience doesn't puzzle over the graphic while you're speaking. Figure F-5 and Figure F-6 show the necessity of labels.

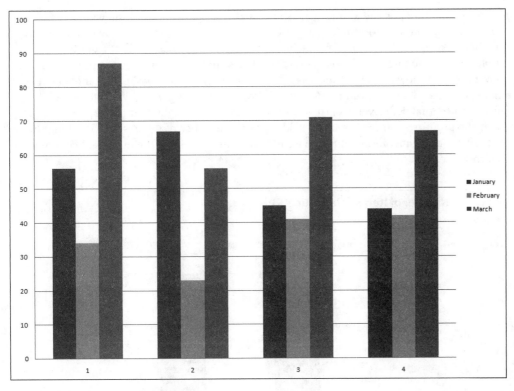

FIGURE F-5 Graphic without labels

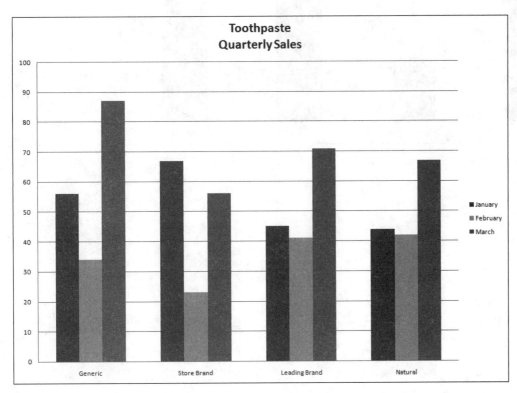

FIGURE F-6 Graphic with labels

In Figure F-5, neither the graphic nor the x- and y-axes are labeled. Do the amounts shown correspond to units or dollars? What elements are represented by each bar? In contrast, Figure F-6 provides a comprehensive snapshot of the business operation, which would support a talk rather than distract from it.

Another common pitfall of producing visual aids is creating charts that have a misleading premise. For example, suppose you want to show how sales have increased and contributed to a growth in net income. If you simply graph the number of items sold in a given month, as displayed in Figure F-7, the visual may not give your audience any sense of the actual dollar value of those items. Therefore, it might be more appropriate (and more revealing) to graph the profit margin for the items sold times the number of items sold. Graphing the profit margin would give a more accurate picture of which item(s) are contributing to the increased net income. That graph is displayed in Figure F-8.

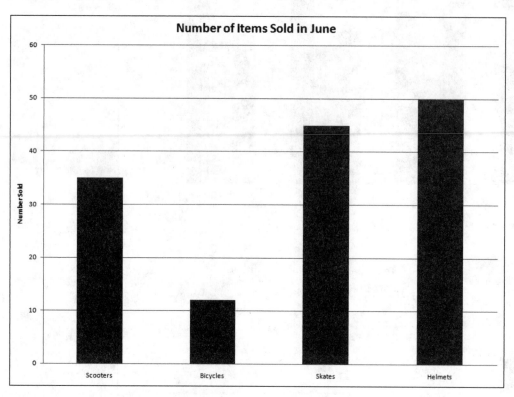

FIGURE F-7 Graph: number of items sold

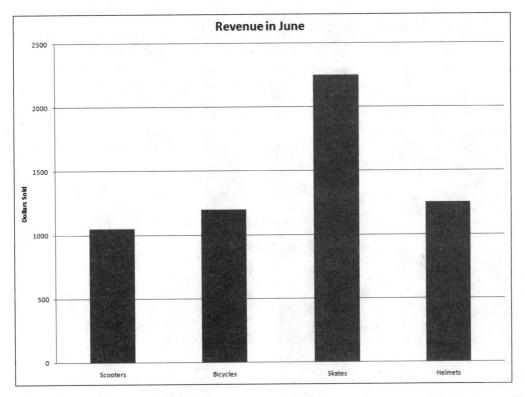

FIGURE F-8   Graph: profit of items sold

Something else you want to avoid is putting too much data in a single comparative chart. Here is an example: Assume you want to compare monthly mortgage payments for two loans with different interest rates and timeframes. You have a spreadsheet that computes the payment data, shown in Figure F-9.

	A	B	C	D	E	F	G
1	**Calculation of Monthly Payment**						
2	Rate	6.00%	6.10%	6.20%	6.30%	6.40%	6.50%
3	Amount	$ 100,000	$ 100,000	$ 100,000	$ 100,000	$ 100,000	$ 100,000
4	Payment (360 payments)	$ 599	$ 605	$ 612	$ 618	$ 625	$ 632
5	Payment (180 payments)	$ 843	$ 849	$ 854	$ 860	$ 865	$ 871
6	Amount	$ 150,000	$ 150,000	$ 150,000	$ 150,000	$ 150,000	$ 150,000
7	Payment (360 payments)	$ 899	$ 908	$ 918	$ 928	$ 938	$ 948
8	Payment (180 payments)	$ 1,265	$ 1,273	$ 1,282	$ 1,290	$ 1,298	$ 1,306

FIGURE F-9   Calculation of monthly payment

In Excel, it is possible to capture all of that information in a single chart, such as the one shown in Figure F-10.

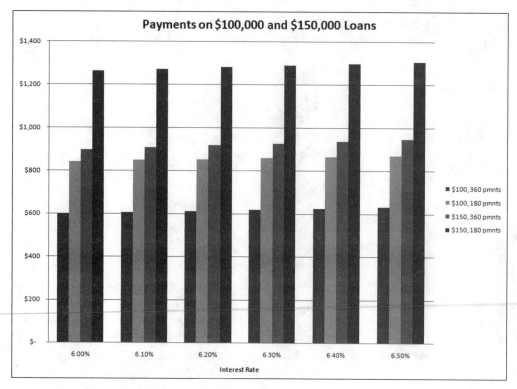

FIGURE F-10 Too much information in one chart

Note, however, that the chart contains a great deal of information. Most readers would probably appreciate your breaking down the information. For example, they would probably find the data easier to understand if you made one chart for the $100,000 loan and another chart for the $150,000 loan. The chart for the $100,000 loan would look like the one shown in Figure F-11.

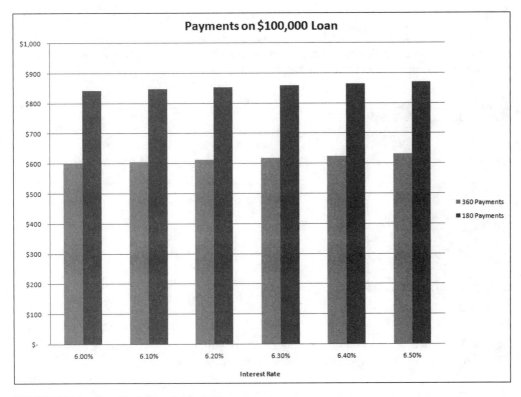

FIGURE F-11   Good balance of information

A similar chart could be made for the $150,000 loan. The charts could then be augmented by text that summarizes the main differences between the payments for each loan. In that fashion, the reader is led step-by-step through the data analysis.

You might want to use the Chart Wizard in Excel, but be aware that the charting functions can be tricky to use, especially with sophisticated charting. Some tweaking of the resultant chart is often necessary. Your instructor might be able to provide specific directions for your individual charts.

Making a Pivot Table in Excel

Suppose you have data for a company's sales transactions by month, by salesperson, and by amount for each product type. You would like to display each salesperson's total sales by type of product sold and by month. You can use a pivot table in Excel to tabulate that summary data. A pivot table is built around one or more dimensions and thus can summarize large amounts of data.

Figure F-12 shows total sales cross-tabulated by salesperson and by month. The following discussion will explain how to create a pivot chart from that data.

	A	B	C	D	E
1	**Name**	**Product**	**January**	**February**	**March**
2	Jones	Product 1	30,000	35,000	40,000
3	Jones	Product 2	33,000	34,000	45,000
4	Jones	Product 3	24,000	30,000	42,000
5	Smith	Product 1	40,000	38,000	36,000
6	Smith	Product 2	41,000	37,000	38,000
7	Smith	Product 3	39,000	50,000	33,000
8	Bonds	Product 1	25,000	26,000	25,000
9	Bonds	Product 2	22,000	25,000	24,000
10	Bonds	Product 3	19,000	20,000	19,000
11	Ruth	Product 1	44,000	42,000	33,000
12	Ruth	Product 2	45,000	40,000	30,000
13	Ruth	Product 3	50,000	52,000	35,000

FIGURE F-12   Excel spreadsheet data

You can create pivot tables (and many other kinds of tables) with the Excel PivotTable tool. To create a pivot table from the data in Figure F-12, follow these steps:

1. Starting in the spreadsheet in Figure F-12, go to the Insert tab; in the Tables group, choose PivotTable. You will see the screen shown in Figure F-13.

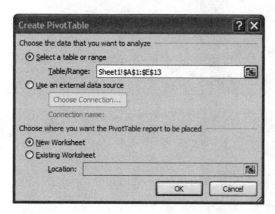

FIGURE F-13 Create pivot table

2. Make sure New Worksheet is checked under Choose where you want the PivotTable report to be placed. Click **OK**. You will then see the screen shown in Figure F-14.

FIGURE F-14 PivotTable design screen

The data range's column headings are shown in the PivotTable Field List on the right side of the screen. From there, you can click and drag column headings into the Row, Column, and Data areas that appear in the spreadsheet.

3. Assume you want to see the total sales by product for each salesperson. You would drag the Name field to the Drop Column Fields Here area in the spreadsheet. After doing that, you should see the result shown in Figure F-15.

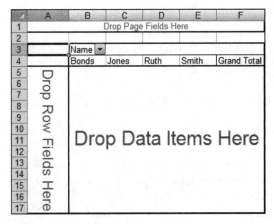

FIGURE F-15   Column fields

4. Next, take the Product field and drag it to the Drop Row Fields Here area. You should see the result shown in Figure F-16.

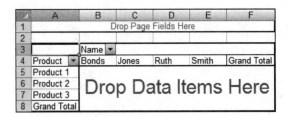

FIGURE F-16   Row fields

5. Finally, take the month fields (January, February, and March) and drag them individually to the Drop Data Items Here area to produce the finalized pivot table; you should see the result shown in Figure F-17.

Product	Data	Bonds	Jones	Ruth	Smith	Grand Total
		Name				
Product 1	Sum of January	25000	30000	44000	40000	139000
	Sum of February	26000	35000	42000	38000	141000
	Sum of March	25000	40000	33000	36000	134000
Product 2	Sum of January	22000	33000	45000	41000	141000
	Sum of February	25000	34000	40000	37000	136000
	Sum of March	24000	45000	30000	38000	137000
Product 3	Sum of January	19000	24000	50000	39000	132000
	Sum of February	20000	30000	52000	50000	152000
	Sum of March	19000	42000	35000	33000	129000
Total Sum of January		66000	87000	139000	120000	412000
Total Sum of February		71000	99000	134000	125000	429000
Total Sum of March		68000	127000	98000	107000	400000

FIGURE F-17   Data items

By default, Excel adds all of the sales for each salesperson by month for each individual product. At the bottom of the pivot table, Excel also shows the total sales for each month for all products.

Creating PowerPoint Presentations

PowerPoint presentations are easy to create. Simply open the application and use the appropriate slide layout for a title slide, a slide containing a bulleted list, a picture, a graphic, etc. When choosing a design template (the background color, the font color and size, and the fill-in colors for all slides in your presentation), keep these guidelines in mind:

- Avoid using pastel background colors. Dark backgrounds such as blue, black, and purple work well on overhead projection systems.
- If your projection area is small or your audience is large, consider using boldfaced type for all of your text to make it more visible.
- Use transition slides to keep your talk lively. A variety of styles are available for use in PowerPoint. Common transitions include dissolves and wipes. Avoid wild transitions such as swirling letters; they will distract your audience from your presentation.
- Use Custom Animation effects when you do not want your audience to see the entire slide all at once. When you use an Entrance effect on each bullet point on a slide, the bullets come up one at a time when you click the mouse or the right arrow. This Custom Animation effect lets you control the visual and explain the elements as you go along. These types of Custom Animation effects can be incorporated and managed under the Custom Animation screen, as shown in Figure F-18.

FIGURE F-18   Custom Animation screen

- Consider creating PowerPoint slides that have a section for your notes. You can print the notes from the Print dialog box by choosing Notes Pages from the Print what drop-down menu, as shown in Figure F-19. Each slide is printed half size, with your notes appearing underneath each slide, as shown in Figure F-20.

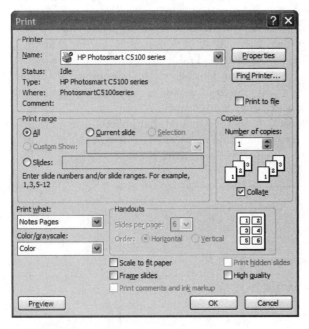

FIGURE F-19   Printing notes page

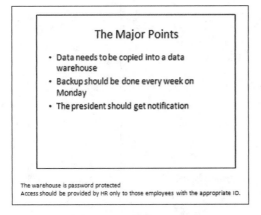

FIGURE F-20   Sample notes page

- You should check your presentation on an overhead. What looks good on your computer screen might not be readable on an overhead screen.

Using Visual Aids Effectively

Make sure you've chosen the visual aids that will work most effectively. Also make sure you have enough—but not too many—visual aids. How many is too many? The amount of time you have to speak will determine the number of visual aids you should use, as will your audience. For example, if you will be addressing a group of teenage summer helpers, you might want to use more visual effects than if you were making a presentation to a board of directors. Remember to use visual aids to enhance your talk, not to replace it.

Review each visual aid you've created to make sure it meets the following criteria:

- The size of the visual aid is large enough so that everyone in the audience can see it clearly and read any labels.

- The visual aid is accurate; for example, the graphics are not misleading, and there are no typos or misspelled words.
- The content of the visual aid is relevant to the key points of your presentation.
- The visual aid doesn't distract the audience from your message. Often when creating Power-Point slides, speakers get carried away with the visual effects; for example, they use spiraling text and other jarring effects. Keep your visuals professional-looking.
- A visual aid should look good in the presentation environment. If possible, try using your visual aid in the presentation environment before you actually deliver your presentation. For example, if you'll be using PowerPoint, try out your slides on an overhead projector in the room in which you'll be showing the slides. As previously mentioned, what looks good on your computer screen might not look good on the overhead projector when viewed from a distance of, say, 20 feet.
- All numbers should be rounded unless decimals or pennies are crucial.
- Slides should not look too busy or crowded. Most experts recommend that bulleted lists not contain more than four or five lines. You want to avoid using too many labels. For an example of a slide that is too busy and, therefore, likely to be ineffective, see Figure F-21.

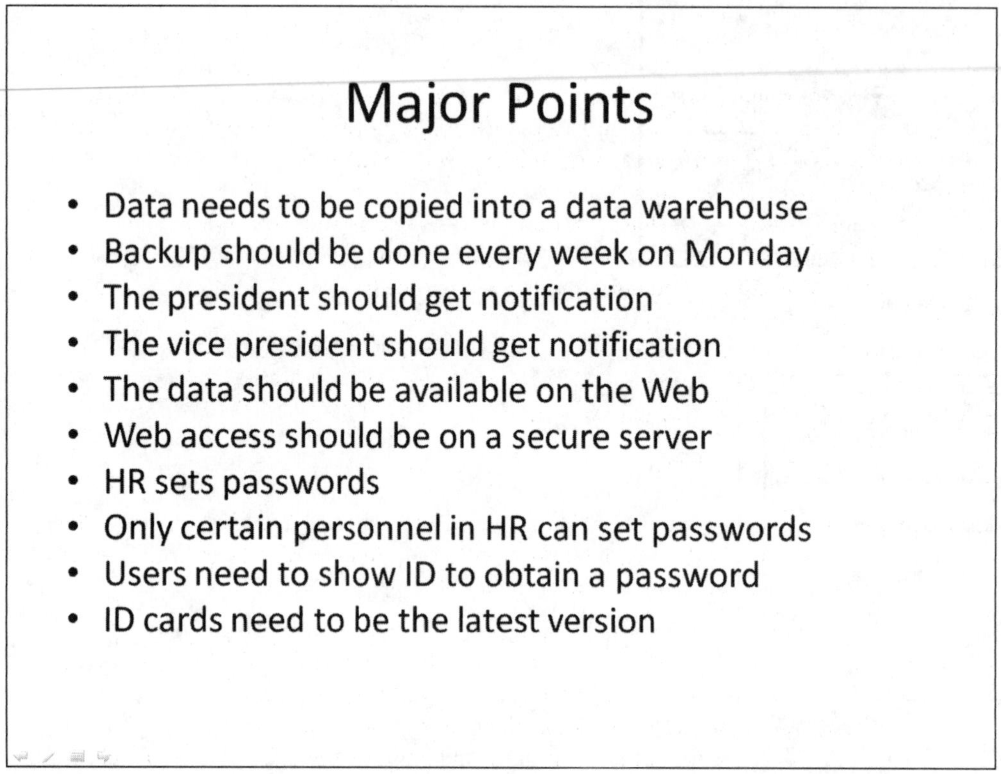

Major Points

- Data needs to be copied into a data warehouse
- Backup should be done every week on Monday
- The president should get notification
- The vice president should get notification
- The data should be available on the Web
- Web access should be on a secure server
- HR sets passwords
- Only certain personnel in HR can set passwords
- Users need to show ID to obtain a password
- ID cards need to be the latest version

FIGURE F-21   Busy slide

PRACTICING YOUR DELIVERY

Surveys indicate that for most people, public speaking is their greatest fear. However, fear or nervousness can be a positive factor. It can channel your energy into doing a good job. Remember that an audience is not likely to perceive you as being nervous unless you fidget or your voice cracks. Audience members want to hear the content of your talk, so think about them and their interests—not about how you feel.

The presentations you give for the cases in this textbook will be in a classroom setting with 20 to 40 students. Ask yourself this question: Am I afraid when I talk to just one or two of my classmates? The answer is probably no. Therefore, you should think of your presentation as an extended conversation with several of your classmates. Let your gaze shift from person to person and make eye contact with various people. As your gaze drifts around the room, say to yourself, I'm speaking to one person. As you become more experienced in speaking before a group, you will be able to let your gaze move naturally from one audience member to another.

Tips for Practicing Your Delivery

Giving an effective presentation is not—and should not—be like reading a report to an audience. Rather, it requires that you rehearse your message well enough so you can present it naturally and confidently and in tandem with well-chosen visual aids. Therefore, you must allow sufficient time to practice your delivery before you give your presentation. Here are some tips to help you hone the effectiveness of your delivery:

- Practice your presentation several times and use your visual aids when you practice.
- Show visual aids at the right time and only at the right time. A visual aid should not be shown too soon or too late. In your speaker's notes, you might include cues for when to show each visual aid.
- Maintain eye and voice contact with the audience when using the visual aid. Don't look at the screen or turn your back on the audience.
- Refer to your visual aids in your talk and with hand gestures. Don't ignore your own visual aid.
- Keep in mind that your visual aids should support your presentation, not *be* the presentation. In other words, don't include everything you are going to say on each slide. Use visual aids to illustrate key points and statistics and fill in the rest of the content with your talk.
- Check your time. Are you within the time limit?
- Use numbers effectively. When speaking, use rounded numbers; otherwise, you'll sound like a computer. Also make numbers as meaningful as possible. For example, instead of saying "in 84.7 percent of cases," say, "in five out of six cases."
- Don't "reach" to interpret the output of statistical modeling. For example, suppose you have input many variables into an Excel model. You might be able to point out a trend, but you might not be able to say with certainty that if a company employs the inputs in the same combination that you used them, the firm will get the same results.
- Record and then evaluate yourself. If that is not possible, have a friend listen to you and evaluate your style. Are you speaking down to your audience? Is your voice unnaturally high-pitched from fear? Are you speaking clearly and distinctly? Is your voice free of distractions, such as *um*, *you know*, *uh*, *so*, and *well*?
- If you use a pointer, either a laser pointer or a wand, be careful. Make sure you don't accidentally direct a laser pointer toward someone's face—you'll temporarily blind the person. If you're using a wand, don't swing it around or play with it.

Handling Questions

Fielding questions from an audience can be an unpredictable experience, because you can't anticipate all of the questions you might be asked. When answering questions from an audience, *treat everyone with courtesy and respect*. Use the following strategies to handle questions:

- Try to anticipate as many questions as possible and prepare answers in advance. Remember that you can gather much of the information to prepare those answers while you draft your presentation. Also, if you have a slide that illustrates a key point but doesn't quite fit in your presentation, save it; someone might have a question that the slide will answer.
- Mention at the beginning of the talk that you will take questions at the end of your presentation. That should prevent people from interrupting you. If someone tries to interrupt, smile and say that you'll be happy to answer all questions when you're finished or that the next graphic will answer the question. If, however, the person doing the interrupting is the CEO of your company, you should answer the question on the spot.
- When answering a question, repeat the question if you have *any* doubt that the entire audience might not have heard it. Then deliver the answer to the whole audience, not just the person who asked the question.
- Strive to be informative, not persuasive. In other words, use facts to answer questions. For instance, if someone asks your opinion about a given outcome, you might show an Excel slide that displays the Solver's output; then you can use that data as the basis for answering the question.
- If you don't know the answer to a question, don't try to fake it. For instance, suppose someone asks you a question about the Scenario Manager that you can't answer. Be honest. Say, "That is

an excellent question; but unfortunately, I don't know the answer." For the classroom presentations you will be delivering as part of this course, you might ask your instructor whether he or she can answer the question. In a professional setting, you can say that you'll research the answer and e-mail the results to the person who asked the question.

- Signal when you are finished. You might say, "I have time for one more question." Wrap up the talk yourself.

Handling a ìProblemî Audience

A "problem" audience or a heckler is every presenter's nightmare. Fortunately, such experiences are rare. If someone is rude to you or challenges you in a hostile manner, remain cool, be professional, and rely on facts. Know that the rest of the audience sympathizes with your plight and admires your self-control.

The problem you will most likely encounter is a question from an audience member who lacks technical expertise. For instance, suppose you explained how to input data into an Access form but someone didn't understand your explanation. In that instance, ask the questioner what part of the explanation was confusing. If you can answer the question briefly, do so. If your answer to the questioner begins to turn into a time-consuming dialogue, offer to give the person a one-on-one explanation after the presentation.

Another common problem is someone who asks you a question that you've already answered. The best solution is to answer the question as briefly as possible using different words (just in case the way in which you explained something confused the person). If the person persists in asking questions that have obvious answers, the person either is clueless or is trying to heckle you. In that case, you might ask the audience, "Who in the audience would like to answer that question?" The person asking the question should get the hint.

PRESENTATION TOOLKIT

You can use the form in Figure F-22 for preparation, the form in Figure F-23 for evaluation of Access presentations, and the form in Figure F-24 for evaluation of Excel presentations.

Preparation Checklist

Facilities and Equipment

- ☐ The room contains the equipment that I need.
- ☐ The equipment works and I've tested it with my visual aids.
- ☐ Outlets and electrical cords are available and sufficient.
- ☐ All the chairs are aligned so that everyone can see me and hear me.
- ☐ Everyone will be able to see my visual aids.
- ☐ The lights can be dimmed when/if needed.
- ☐ Sufficient light will be available so I can read my notes when the lights are dimmed.

Presentation Materials

- ☐ My notes are available, and I can read them while standing up.
- ☐ My visual aids are assembled in the order that I'll use them.
- ☐ A laser pointer or a wand will be available if needed.

Self

- ☐ I've practiced my delivery.
- ☐ I am comfortable with my presentation and visual aids.
- ☐ I am prepared to answer questions.
- ☐ I can dress appropriate for the situation.

FIGURE F-22   Preparation checklist

Evaluating Access Presentations

Course: _____ Speaker: _____ Date: _____

Rate the presentaton by these criteria:
4=Outstanding   3=Good   2=Adequate   1=Needs Improvement
N/A=Not Applicable

Content

_____ The presentation contained a brief and effective introduction.

_____ Main ideas were easy to follow and understand.

_____ Explanation of database design was clear and logical.

_____ Explanation of using the form was easy to understand.

_____ Explanation of running the queries and their output was clear.

_____ Explanation of the report was clear, logical, and useful.

_____ Additional recommendations for database use were helpful.

_____ Visuals were appropriate for the audience and the task.

_____ Visuals were understandable, visible, and correct.

_____ The conclusion was satisfying and gave a sense of closure.

Delivery

_____ Was poised, confident, and in control of the audience

_____ Made eye contact

_____ Spoke clearly, distinctly, and naturally

_____ Avoided using slang and poor grammar

_____ Avoided distracting mannerisms

_____ Employed natural gestures

_____ Used visual aids with ease

_____ Was courteous and professional when answering questions

_____ Did not exceed time limit

Submitted by: _____

FIGURE F-23   Form for evaluation of Access presentations

Evaluating Excel Presentations

Course: _____ Speaker: _____ Date: _____

Rate the presentaton by these criteria:
4=Outstanding   3=Good   2=Adequate   1=Needs Improvement
N/A=Not Applicable

Content

_____ The presentation contained a brief and effective introduction.

_____ The explanation of assumptions and goals was clear and logical.

_____ The explanation of software output was logically organized.

_____ The explanation of software output was thorough.

_____ Effective transitions linked main ideas.

_____ Solid facts supported final recommendations.

_____ Visuals were appropriate for the audience and the task.

_____ Visuals were understandable, visible, and correct.

_____ The conclusion was satisfying and gave a sense of closure.

Delivery

_____ Was poised, confident, and in control of the audience

_____ Made eye contact

_____ Spoke clearly, distinctly, and naturally

_____ Avoided using slang and poor grammar

_____ Avoided distracting mannerisms

_____ Employed natural gestures

_____ Used visual aids with ease

_____ Was courteous and professional when answering questions

_____ Did not exceed time limit

Submitted by: _____

FIGURE F-24   Form for evaluation of Excel presentations

INDEX

U

V